PHILOSOPHY OF THE SOCIAL SCIENCES

Other interview books from Automatic Press ♦ VIP

Formal Philosophy
edited by Vincent F. Hendricks & John Symons
November 2005

Masses of Formal Philosophy
edited by Vincent F. Hendricks & John Symons
October 2006

Political Questions: 5 Questions for Political Philosophers
edited by Morten Ebbe Juul Nielsen
December 2006

Philosophy of Technology: 5 Questions
edited by Jan-Kyrre Berg Olsen & Evan Selinger
February 2007

Game Theory: 5 Questions
edited by Vincent F. Hendricks & Pelle Guldborg Hansen
April 2007

Legal Philosophy: 5 Questions
edited by Morten Ebbe Juul Nielsen
October 2007

Philosophy of Mathematics: 5 Questions
edited by Vincent F. Hendricks & Hannes Leitgeb
January 2008

Philosophy of Computing and Information: 5 Questions
edited by Luciano Floridi
Sepetmber 2008

Epistemology: 5 Questions
edited by Vincent F. Hendricks & Duncan Pritchard
September 2008

Complexity: 5 Questions
edited by Carlos Gershenson
September 2008

Probability and Statistics: 5 Questions
edited by Alan Hájek & Vincent F. Hendricks
September 2008

See all published and forthcoming books in the 5 Questions series at
www.vince-inc.com/automatic.html

PHILOSOPHY OF THE SOCIAL SCIENCES

5 QUESTIONS

edited by

Diego Ríos

Christoph Schmidt-Petri

Automatic Press ♦ $\frac{V}{I}$P

Automatic Press ♦ $\frac{V}{I}$P

Information on this title: www.vince-inc.com/automatic.html

© Automatic Press / VIP 2008

This publication is in copyright. Subject to statuary exception
and to the provisions of relevant collective licensing agreements,
no reproduction of any part may take place without
the written permission of the publisher.

First published 2008

Printed in the United States of America
and the United Kingdom

ISBN-10 87-92130-12-7   paperback
ISBN-13 978-87-92130-12-9 paperback

The publisher has no responsibilities for
the persistence or accuracy of URLs for external or
third party Internet Web sites referred to in this publication
and does not guarantee that any content on such
Web sites is, or will remain, accurate or appropriate.

Typeset in $\LaTeX 2_\varepsilon$
Photo and graphic design by Vincent F. Hendricks

Contents

Preface	iii
Acknowledgements	v
1 David Bloor	1
2 Raymond Boudon	15
3 Mario Bunge	31
4 Nancy Cartwright	43
5 Margaret Gilbert	47
6 Daniel M. Hausman	57
7 Harold Kincaid	69
8 Daniel Little	79
9 Steven Lukes	89
10 David Papineau	103
11 Philip Pettit	115
12 Alexander Rosenberg	125
13 David-Hillel Ruben	129
14 John R. Searle	139
15 Raimo Tuomela	143
About the Editors	151
Index	154

Preface

Within the general landscape of mainstream philosophy of science, the philosophy of the social sciences is a rather underdeveloped domain. At least in the analytical tradition, the social sciences have received comparatively less attention than the natural sciences. There might be different reasons explaining this situation; some of them will emerge, we hope, in the essays that are published in this volume. One important reason is that philosophers and social scientist have for a long time been quite disconnected from each other. This has changed in the recent past, and indeed the last two decades have shown important signs of genuine communication across disciplines. *Philosophy of the Social Sciences: Five Questions* should be read as a small contribution to these exchanges between philosophers and social scientists, making salient in an informal way the main topics or problems they are concerned with.

But the objective of this volume is not just to work as *rendezvous* for philosophers and social scientists, but also to isolate those problems or issues in the philosophy of the social sciences that, according to some of its most prominent practitioners, are currently – or are likely to become – hot topics in their disciplines.

In order to achieve these objectives, we have been quite flexible in the way we have organized the volume. To begin with, we have attempted to include a wide range of points of views, independently of philosophical persuasion. Although the contributors are still mostly analytical philosophers, social scientists that have a more practical appraisal of their own work are also represented. We have furthermore attempted to cover a number of different social sciences with the intention of providing a general perspective on the field. Last but not least, we have given the contributors a lot of liberty in the way they could write their own papers. Although we suggested five questions that seemed particularly interesting or important to us, the contributors were free to reply to them one by one, rephrase them, write one longer essay, or simply to discuss other issues that they personally consider of paramount importance. In a nutshell: rather than setting ourselves the agenda, we have given the contributors the opportunity to set it by themselves. The outcome of this project is a volume where difficult problems are tackled in a lively way by the main protagonists of the debates.

Independently of the differences of emphasis or style, we think there are some recurrent themes that most of the contributors of these essays mention in their own papers. First, the issues surrounding reductionism: what is the relationship between the social sciences and the other – presumably more basic – disciplines like psychology or biology? Second, the problem of relativism: how should social scientists deal with radically alien forms of behaviour or beliefs? Third, the problem of explanation: how does explanation work in the social sciences? Fourth, the issue of causality: what is the philosophical status of the tools that we have to isolate causal relationships in the social sciences? Last but not least, there are metaphysical issues concerning the proper ontology of the social sciences.

───────────────── ♦ ─────────────────

This book presents these themes in a lively way, and connects them to the lives of the most important thinkers working on them. We believe it should be interesting for anyone who cares about the social sciences.

Diego Ríos & Christoph Schmidt-Petri
May 2008

Acknowledgements

We thank Julian Reiss and Vincent F. Hendricks for help and advice in preparing this volume. We would like to thank Claus Festersen and Rasmus Rendsvig when encountering LaTeX-related problems. And finally we would like to thank our publisher Automatic Press ♦ $\frac{\lor}{\mid}$P, in particular senior publishing editor V.J. Menshy, for continuing to take on these 'rather unusual academic' projects.

Diego Ríos & Christoph Schmidt-Petri
May 2008

1

David Bloor

Science Studies Unit
University of Edinburgh, UK

How did you get interested in the philosophical aspects of the social sciences?

I was an undergraduate at the University of Keele between 1960 and 1964. As part of a laudable policy to offset over-specialisation the regulations required students to proceed to a joint honours degree[1]. They were to divide their time equally between two departments. I chose mathematics and philosophy. This is how I came to read Wittgenstein's *Philosophical Investigations*[2] and Hume's *Treatise*[3]. Wittgenstein identified the fundamental role of conventions in all human knowledge, though he never explained what a convention was. Hume had already given the definitive account of conventions in Book III of the *Treatise*, so the two approaches fitted together. Both of these thinkers were routinely identified as 'philosophers' but it was clear that neither of them had any patience with the *a priori* procedures that were, and are, characteristic of the academic pursuit that went under this name. For Wittgenstein philosophical problems arise when "language *goes on holiday*", that is, when language is used in an irresponsible fashion disconnected from the real business of life[4]. For Hume the written productions of philosophers, in as far as they did not deal with matters of fact or quantity and number, went under the name

---

[1] Keele was a new, experimental university, the brainchild of the Oxford idealist philosopher A.D. Lindsay. W.B. Gallie, *A New University: A.D. Lindsay and the Keele Experiment*, London, Chatto and Windus, 1960.
[2] L. Wittgenstein, *Philosophical Investigations*, Oxford, Blackwell, 1953.
[3] D. Hume, *A Treatise of Human Nature*, (ed. L.A. Selby-Bigge), Oxford, Clarendon Press, 1888.
[4] L. Wittgenstein, *Philosophical Investigations*, Oxford, Blackwell, 1953, paragraph 38.

of 'metaphysics' and represented "nothing but sophistry and illusion". They should, he said, be thrown onto the bonfire[5]. More positively, Wittgenstein said: "don't think, but look!", while Hume said his aim was "to introduce the experimental method of reasoning into moral subjects"[6]. Of the two, it was Hume's scientific orientation, rather than Wittgenstein's more equivocal attitude to science, that impressed me most. But both of them clearly wanted to give their enquiries an empirical basis. This is what I took from these writers when I was a student and it is what I believe as strongly today. I can therefore only speak of having an interest in the 'philosophical' aspects of the social sciences in a negative way. It would be more accurate to say that these works led me away from philosophy rather than towards it.

It will be clear that when I use the word 'philosophy' I mean the academic activity as I was exposed to it forty years ago. Academic fashions change, especially in non-empirical fields and, no doubt, in many respects departments of philosophy are now different. But, judging by the debates I am currently having with philosophers, I suspect that they are not all that different. The professor of philosophy at Keele was A.G.N. Flew, a forceful, Oxford-trained, ordinary-language philosopher. Apart from the philosophical classics, the staple diet in the department was provided by the exponents of the Oxford approach in its various forms. We cut our teeth on the writings of Ryle, Austin and Strawson. Flew had gathered the core papers into three collections, *Essays in Conceptual Analysis*, and *Logic and Language, Series I* and *Series II*[7]. There was much of educational value in the close study and dissection of these papers. I enjoyed the activity but could not rid myself of the uneasy lesson I was learning from Wittgenstein and Hume. What exactly was 'conceptual analysis'? How could it be done on the basis of little more than informal, linguistic intuition? Where

---

[5] D. Hume, *An Enquiry Concerning Human Understanding*, (ed. L.A. Selby-Bigge), Oxford, Clarendon Press, 2nd ed. 1902, p.165. (For a modern reader the act of burning books will bring to mind Dr. Goebbels. The background to Hume's remarks was very different. His readers will have understood his position in relation to Savonarola's bonfire of the vanities.)

[6] The quotation from Wittgenstein comes from *Investigations* paragraph 66. The quotation from Hume is from the title page of the *Treatise*.

[7] A.G.N. Flew (ed.) *Logic and Language (First Series)* Oxford, Blackwell, 1951; A.G.N. Flew (ed.) *Logic and Language (Second Series)* Oxford, Blackwell, 1953; Antony Flew, *Essays in Conceptual Analysis*, London, Macmillan, 1960; Antony Flew, *Hume's Philosophy of Belief. A Study of his First Inquiry*, London, Routledge and Kegan Paul, 1961.

did this leave the natural sciences and their history of theoretical change? Ryle spoke of philosophy as the study of the 'logical geography' of concepts but the resulting activity did not conform to the metaphor. Where was the systematic and objective study of the terrain? Wittgenstein and Hume were amongst the officially listed heroes of the analytical movement but there seemed to be a conspiracy of silence about their main message.

Which social sciences do you consider particularly interesting or challenging from a philosophical point of view?

After Keele I went to Trinity College, Cambridge for post-graduate work. I was encouraged in this move by the philosopher Alan Ryan who was then a young lecturer at Keele. In 1964 I began research in the history and philosophy of science supervised by Mary Hesse. I was, and still am, deeply impressed by her work on inductive inference and its relation to the role played by models and metaphors in the construction of scientific theories[8]. But though this was a different, and healthier, sort of philosophy I felt disillusioned with philosophy in all its forms. I decided that experimental psychology must be the proper way to analyse concepts and to tackle many of the problems that had hitherto been called 'philosophical'. With Mary Hesse's kind help, and agreement from Trinity, I transferred to the Department of Psychology and prepared myself for the psychology papers of Part II of the Biological Sciences Tripos.

Experimental psychology in Cambridge was tough-minded and focussed on the psychology of perception, memory, learning and sensory-motor skills[9]. I was fortunate to be supervised by A.J. Watson, who had himself made the transition from philosophy to experimental psychology. Every week I would read him an essay on a psychological topic he had chosen and every week he would demolish my account of the problem and my account of the current

---

[8] Mary B. Hesse, *Forces and Fields. A Study of Action at a Distance in the History of Physics*, London, Nelson, 1961; Mary Hesse, *The Structure of Scientific Inference*, London, Macmillan, 1974; Mary Hesse, *Revolutions and Reconstructions in the Philosophy of Science*, Brighton, Harvester Press, 1980.

[9] See D.E. Broadbent, *Perception and Communication*, Oxford, Pergamon, 1958; A.T. Welford, *Fundamentals of Skill*, London, Methuen, 1968; D.E. Broadbent, *In Defence of Empirical Psychology*, London, Methuen, 1973; R. Gregory, *Concepts and Mechanisms of Perception*, London, Duckworth, 1974.

state of understanding. It was as exciting as it was dismaying. I felt that I had to learn an entirely new way of thinking. Although I was trying to cast philosophy aside I was still thinking in a philosophical, backwards-oriented, way. Watson showed me that to explore the content of a concept one should not go backwards, to definitions, but forwards, and use the concept to make predictions. This was not simply a case of drawing out routine or pre-existing 'implications' but of putting the idea to work in new ways and using it to explore novel, experimentally contrived, circumstances. Content had to be created rather than extracted. This really did begin to look like Hume's introduction of the experimental method into the moral sciences. It also gave me a deeper appreciation of the way Hume had himself used the rich and concrete material of legal debate in his own work.

Although the Cambridge Psychological Laboratory had originally been founded by W.H.R. Rivers and C.S. Myers, it was Sir Frederic Bartlett F.R.S who stamped his personality on the enterprise in the inter-war years and gave it its modern form[10]. Bartlett's classic book *Remembering. A Study in Experimental and Social Psychology* (1932) was still required reading when I was a student[11]. *Remembering* is a deceptively simple work, framed in qualitative and common-sense terms, but Bartlett was a powerful and subtle thinker. Unusually, for a laboratory study of memory at that time, Bartlett did not ask his subjects to recall meaningless nonsense syllables. The accepted justification for using such material was that it stripped away irrelevant variables and exposed the essential mechanics of the memory, uncontaminated by prior knowledge and idiosyncratic associations. Bartlett rejected this line of reasoning. He insisted that the complexity of a response was determined by the complexity of the responding organism, not the complexity of the task or stimulus. A complex, human organism would respond in complex ways even if given a 'simple' task. Far better, Bartlett argued, to confront this complexity head-on. He therefore used real-life, concrete material. Rather than nonsense syllables he used a folk story, in fact, a folk story drawn from a culture quite different to that of his experimental subjects. Bartlett asked his Cambridge subjects to read a North-American

---

[10] D. E. Broadbent, Frederic Charles Bartlett, 1886-1969, *Biographical Memoirs of Fellows of the Royal Society*, vol. 16, 1970, 1-13; A. Saito (ed.) *Bartlett, Culture and Cognition*, London, Psychology Press, 2000.

[11] F.C. Bartlett, *Remembering. A Study in Experimental and Social Psychology*, Cambridge, C.U.P., 1932.

Indian folk story and then write it down as best they could some time later. After a few days he asked them to write it down again, and then again after a further lapse, and so on. He compared the different versions to see how they had changed. He discovered the gradual but systematic 'conventionalisation' of the story as its original features were changed into a story closer to the norms and pre-occupations of the subject's own culture. A second experiment produced a similar result though here each subject's remembered version became the starting point for the next subject, in the manner of a rumour passing from person to person. On the basis of such experiments Bartlett was able to provide detailed evidence of both the creative and the socially structured character of cognitive processes.

In Chapter XVI of *Remembering* he extended his approach from cultural material like folk-stories to activity in science and technology. In the inter-war years, Bartlett had been involved as a psychologist in the development of sound locators for the night-detection of aircraft. These devices had been introduced during WWI and Bartlett used them to illustrate the significance of his conclusions. He argued that different national groups had developed subtly different forms of locator technology. The circulation of ideas within each group conventionalised the device in different and characteristic ways. The creative and collective processes involved in the development were those uncovered by his experiments on memory. Significantly he called his account of cognition "social constructiveness".[12]

When present day philosophers sneer at current work by social constructivists they never fail to represent it as 'fashionable' and 'trendy'. But the terminology and the substance of the approach goes back to 1930s Cambridge and Bartlett's laboratory. It is difficult to hang the label 'trendy' onto the forthright Sir Frederic. Nor could there be anything more challenging, from a philosophical point of view, than his line of work. Here was an exemplar of how to carry forward the study of knowledge in an empirical rather than an *a priori* way. What role was there for departments of philosophy when there were departments of psychology, along with anthropology, sociology and history, that had set in train a naturalistic account of epistemology? Philosophy, I concluded,

---

[12] For a discussion of Bartlett's sound-locator example, see: D. Bloor, "Whatever happened to 'social constructiveness'?" in A. Saito (ed.) *Bartlett, Culture and Cognition*, London, Psychology Press, 2000, 194-215.

was just armchair psychology. It was defunct and the departments devoted to it should be disbanded.

How do you conceive the relation between the social sciences and the natural sciences?

The relation between the natural and the social sciences is not a relation between two logical essences but between two social institutions. Since we construct, rather than discover, our own institutions then it is up to us to make that relation what we want it to be. Of course, we do not construct institutions as if they are unconstrained fantasies. We must work with recalcitrant material and take human nature, history, and the empirical world as we find it. Nevertheless there is no unique way to live in the world and no unique way to construct the relationship in question. We can emphasise differences between the natural and social sciences or we can emphasise the similarities. I want to stress the similarities.

Works such as Bartlett's *Remembering*, which expressed a straight-forward, empirical curiosity, provide the model. Cognition is a human phenomenon and humans are to be understood against the backdrop of their biological nature. The process of remembering is to be understood in an evolutionary way, so mistakes are as important as successes. They both play a role in the overall story. To say that (by definition) you have only truly 'remembered' X if X really happened may be good conceptual analysis but it is bad experimental psychology. The memory is not to be understood merely as a faculty of accurate reproduction but as the means by which we bring the past to bear on the present. How that is done, and what role is played by accuracy and inaccuracy, is a matter to be determined by observation not definition. What others saw, in a narrow and negative light, as inaccuracies of recall, Bartlett came to see in a positive light as creative acts which must have a biological and social function. His curiosity, like that of any other natural scientist, but unlike that of most philosophers, was not bifurcated by superficial, prior evaluation. He adopted what would now be called a 'symmetrical' approach.

Can a naturalistic orientation of this kind be carried through to the point where there is a single, unified science in which the same concepts are used to describe machines and people? Isn't there a logical gulf? Doesn't the presence of intentionality in the human sciences and its absence in the natural sciences point to a division that can never be bridged? The answer is that no one knows how far the similarities can be taken. The lesson to be learned from scientists such as Bartlett is that it pays to be relaxed about these logical relationships. Concepts change and logical relationships are re-configured. The investigator starts with familiar, common-sense categories and modifies them in response to the experimental results. Theories are constructed and new concepts are introduced by the use of analogy and metaphor. In this way non-intentional concepts and mechanistic models gradually infiltrate the discourse.

Good philosophers and good scientists both know this. Hesse knew explicitly what Bartlett knew implicitly. Her account of the theoretical re-description of empirical phenomena shows the positive role of these logical shifts in the course of scientific enquiry. Calling light a 'wave' required the physicist to modify the previous concept of 'light'. The metaphorical displacement of concepts of this kind can create an initial sense of absurdity until familiarity and entrenchment allows the new pattern of description to be taken as literal[13]. What physicists do with the concept of light, psychologists do with the concepts of remembering, perception, action, intention, mistake and responsibility. Bartlett himself, in his later work, did this when he construed 'fatigue', not in common sense terms as 'being tired', nor in physiological terms as the accumulation of toxins in muscle tissue, but as a form of limited information-processing capacity. Philosophers who chart the 'logical geography' of concepts have difficulty with this characteristic feature of scientific practice. They are prone to identify 'category errors' in what is, in fact, the very stuff of creative, scientific thinking[14]. It is therefore futile to try to map out in advance the logical relationships that will emerge in some future state of science or some future common-sense, and this applies to the logical relationships between the concepts of the future social and natural sciences. But though we can have no clear idea what the future might look like, it is still desirable to push the natural and social sciences closer together. In practice this means: always look for

---

[13] Mary Hesse, "The Explanatory Function of Metaphor ", reprinted in Mary Hesse, *Revolutions and Reconstructions in the Philosophy of Science*, Brighton, Harvester Press, 1980, 111-124.

[14] D. Bloor, "The Dialectics of Metaphor", *Inquiry*, vol. 14. 1971, 430-444; D. Bloor, "Are Philosophers Averse to Science?", in D.O. Edge and J.N. Wolfe (eds) *Meaning and Control. Essays in Social Aspects of Science and Technology*, London, Tavistock, 1973, 1-17.

causes[15].

What is the most important contribution that philosophy has made to the social sciences?

If I were to accept the terms in which the question is asked I would re-iterate what I said above and nominate Hume's analysis of convention as the most important contribution. As part of that analysis Hume also sketched an account of what is today called the problem of collective action and the free-rider problem, (see Part ii, Book III). He effectively showed that a society of rational, calculating, self-interested agents could never produce collective, indivisible goods. For example: they could never co-operate to remove pollution. Given that today's political and economic thinking is saturated with the fantasies of free-market individualism this result could hardly have greater importance. It explodes the myth of the present age and shows the appalling consequences that await us if we act in accordance with the dominant ideology of our times.

I do not, however, accept the terms in which the question is posed. I do not see this important work as a gift made by philosophers to sociologists. Rather, I see Hume as a pioneering sociologist whose work has been selectively appropriated by members of an academic sub-group who are, for the most part, antagonistic to everything for which he stood. If it is insisted, as a verbal point, that Hume's work should be given the label 'philosophical' then I would re-express my position in the following way. It amounts to using the word 'philosophy' to describe certain phases of theory construction in the sciences. The philosophy of sociology then becomes high-level sociology and the philosophy of physics becomes highly general, theoretical physics. This suggest that the best place to do the 'philosophy' of sociology is in the department of sociology and the best place to do the 'philosophy' of physics is in the physics department. In this way high-level theorising will not lose contact with empirical work in the relevant field. By contrast, departmental structures which sever this connection

---

[15] Some students of social and cognitive phenomena, e.g. 'ethnomethodologists', deliberately avoid causal analysis. It is interesting to notice how similar their approach is to that of the Oxford ordinary-language philosophers. This extends to their preferred reading of Wittgenstein. See D. Bloor, "Left and Right Wittgensteinians", in A. Pickering (ed.) *Science as Practice and Culture*, Chicago, University of Chicago Press, 1992, 266-282.

will encourage language and thinking which is permanently "on holiday".

Which topics in the philosophy of social science will, and which should, receive more attention than in the past?

After Cambridge I needed a job. Given my mixture of qualifications, and my opinions, I was fortunate in being able to find employment in a new, interdisciplinary venture. In 1967, at the prompting of the eminent biologist C.H. Waddington, the University of Edinburgh created the Science Studies Unit under the directorship of the former radio-astronomer David Edge[16]. The aim of the Unit, as far as teaching was concerned, was to provide courses in 'Science and Society' to supplement the specialist training of science undergraduates. My colleagues in the new venture were the sociologist Barry Barnes[17] and the historian Steven Shapin[18]. Collectively we arrived at an approach to science in which scientific knowledge was to be analysed as a social achievement and as an element of culture. In other words, we started doing the sociology of knowledge. We did not see this approach as novel; we were simply following tendencies that were already well established within history and sociology. And certainly none of us saw it as an expression of hostility to science. The aim was explanation not criticism, though the work that resulted from the approach was frequently represented as criticism by commentators who could not tell the difference. In fact, the approach was an expression of respect for science. It was meant to be a scientific way of understanding the status and character of science itself.

---

[16] D.O. Edge and M.J. Mulkay, *Astronomy Transformed. The Emergence of Radio Astronomy in Britain*, New York, Wiley, 1976.

[17] As a small but representative sample of publications see: Barry Barnes, *Scientific Knowledge and Sociological Theory*, London, Routledge, 1974; Barry Barnes, *T.S. Kuhn and Social Science*, London, Macmillan, 1982; Barry Barnes, *The Nature of Power*, Cambridge, Polity Press, 1988; Barry Barnes, David Bloor and John Henry, *Scientific Knowledge. A Sociological Analysis*, Chicago, University of Chicago Press, 1996.

[18] Again a small sample must suffice: S. Shapin, "History of Science and its Sociological Reconstructions", *History of Science*, vol. xx, 1982, 157-211; S. Shapin and S. Schaffer, *Leviathan and the Air-Pump. Hobbes, Boyle, and the Experimental Life*, Princeton, Princeton University Press, 1985; S. Shapin, *A Social History of Truth. Civility and Science in Seventeenth-Century England*, Chicago, University of Chicago Press, 1994. S. Shapin, *The Scientific Revolution*, Chicago, University of Chicago Press, 1996.

1. David Bloor

For me the sociology of knowledge was an obvious extension of the causal and naturalistic approach I had learned in the Cambridge Psychological Laboratory. The methodology was the same except that the data now typically came from history rather than experiment[19].

There is one feature of this collective approach that, from the outset, has caused particular controversy, namely, its 'relativism'[20]. What is at issue here? Relativism is frequently presented by critics as if it means 'anything goes'. This cannot be right. The central point behind all forms of relativism must be the denial of absolutism. Relativism is the claim that human cognition possess no properties that justify calling it 'absolute' on any reasonable definition of that term. Human knowledge is real but it cannot to be considered as infallible, eternal, uncaused, or unconditioned. Knowledge is relative to the human condition, the limitations of the human intellect and our historical situation and cultural resources. It is a natural phenomenon enmeshed in the causal nexus, not a supernatural thing. Relativism is an immediate consequence of adopting a naturalistic perspective and hence something demanded by science as we now understand it. Absolute knowledge cannot be conjectural, provisional, inconsistent, partly right and partly wrong—and yet scientific knowledge cannot be anything else.

Once again, the central battle had already been fought and won by Hume. Hume argued that all empirical knowledge is inductive knowledge and inductive knowledge cannot be justified in a non-circular way. Nobody has answered Hume's challenge. No absolute justification for inductive knowledge has ever been produced. All inductive conclusions are relative to our presuppositions, cultural assumptions, limited evidence, search strategies, goals, interests and natural, animal proclivities to generalise. In other words, Hume has already established relativism with respect to empirical knowledge. Every philosopher knows Hume's

---

[19] D. Bloor, *Knowledge and Social Imagery*, (2nd edition) Chicago, University of Chicago Press, 1991; D. Bloor, *Wittgenstein. A Social Theory of Knowledge*, London, Macmillan, 1983; D. Bloor, *Wittgenstein, Rules and Institutions*, London, Routledge, 1997; D. Bloor, Sociology of Scientific Knowledge, in I. Niiniluoto, M. Sintonen, and J. Wolenski (eds), *Handbook of Epistemology*, Dordrecht, Kluwer, 2004, 919-962.

[20] B. Barnes and D. Bloor, "Relativism, Rationalism and the Sociology of Knowledge", in M. Hollis and S. Lukes (eds) *Rationality and Relativism*, Oxford, Blackwell, 1982, 21-47.

result but, once again, the profession, which is overwhelmingly anti-relativist, is in denial.

This is one area that *should* receive more attention, provided it is of the right kind. The intellectual standard of anti-relativist argumentation, particularly that directed against the Edinburgh approach, is scandalous and needs to be put right. Do those who reject relativism have any idea of the problems they confront if they are to justify a truly non-relative status for any aspect of our scientific knowledge? They have got to justify induction, though their problems do not end there[21]. It is little wonder that they fudge the issue. They make life easy for themselves by attacking some specific version of relativism (usually some bizarre version of 'anything goes') and then treat this as representative of all forms of relativism. Again, the critics systematically conflate relativism with a range of quite different doctrines. They then refute or, at least, reject these other doctrines and believe, wrongly, that they have refuted relativism.

Let me illustrate these fudges. It is often assumed that relativism can be rejected by asserting the objectivity of science. Evidence is presented to show that knowledge can properly be called 'objective' and the conclusion is drawn that relativism is false. The conclusion does not follow. It rests on the mistaken assumption that relativism can be equated with subjectivism. Whilst it may be true that subjectivism implies relativism, it is not true that relativism implies subjectivism. An argument based on the appeal to objectivity needs to show that objectivity can only be understood in absolutist terms. Only an absolute form of objectivity can refute relativism and the possibility of absolute objectivity cannot be taken for granted. But this is exactly what the critics do take for granted. Their argument is question-begging. It appears to escape the critics of relativism that it is possible to produce a relativist analysis of the objectivity of knowledge[22].

---

[21] What about Popper's attempt to build an account of science that only uses deductive inferences and hence, apparently, does not depend on induction? Does that help the absolutist? Popper's essential point is that inductive inferences are really unjustified and unjustifiable conjectures. A relativist can accept this point. It is really just another way of stating Hume's position. As such it is no help to the absolutist.

[22] Durkheim showed how to do this. Those who speak of the objective character of knowledge are really describing its social character. See: D. Bloor, "A Sociological Theory of Objectivity", in S.C. Brown (ed.), *Objectivity and Cultural Divergence (Royal Institution of Philosophy Lecture Series, 17)*, Cambridge, Cambridge University Press, 1984, 229-245, For a Durkheim-style

Similarly, critics of relativism often assert (rightly) that science is a rational activity and then conclude (wrongly) that science cannot be relative. But, in a similar way to the previous argument, this presupposes that rationality must be understood in absolute terms and cannot be understood in relative terms. The crucial premise is rarely recognised, let alone defended. Most common of all, critics assume that relativism is a species of philosophical idealism and implies the denial of a mind-independent reality. In the mind of the critic any argument or assertion that there is an independent, material reality therefore passes as a refutation of relativism. Of course, it is no such thing. Relativism is not the same as idealism. You can be, and should be, a materialist and a relativist.

Throughout all these anti-relativist 'arguments' the relativist-absolutist dichotomy is being confused with some other, quite different, dichotomy: the idealist-materialist dichotomy, the subjectivist-objectivist dichotomy and the irrationalist-rationalist dichotomy. These dichotomies all have their own specific jobs of work to do and it is inexcusable to run them together and confuse them with the relativist-absolutist dichotomy. This conflation has become endemic in academic philosophy and is to be found, for example, in both the *Oxford Companion to Philosophy* and the *Cambridge Dictionary of Philosophy* [23]. How this confusion is possible in a discipline whose practitioners pride themselves on verbal exactitude and conceptual clarity is, at first glance, something of a mystery.

The more I see of the current philosophical attacks on relativism, however, the more convinced I am that my youthful instinct was sound and that the source of all this trouble was correctly diagnosed by Hume and Wittgenstein. There is something fundamentally unhealthy about philosophy as a distinct, non-empirical enterprise. There is also something suspect about the way many of its practitioners relate to the natural sciences as the dominant strand in our culture. The trouble starts when philosophers define their field as a form of purely conceptual ex-

---

reading of Popper's theory of objective knowledge see: D. Bloor, "Popper's Mystification of Objective Knowledge", *Science Studies* (now *Social Studies of Science*) vol. 4, 1974, 65-76.

[23] C.A.J. Coady, "Relativism, epistemological", in T. Honderich (ed.) *The Oxford Companion to Philosophy*, Oxford, OUP, 1995, p. 757 and L.P. Pojman, "Relativism", in Robert Audi (ed.) *The Cambridge Dictionary of Philosophy*, Cambridge, Cambridge University Press, 1995, 690-691.

ploration or legislation, devoted to replacing category habits by category disciplines. Leave esoteric matters of fact to the scientists, they say. We are concerned with logical and conceptual relations. Sometimes (as with Oxford ordinary-language philosophy) this amounted to a defence of everyday, common-sense against the unwelcome incursions of science. But, more recently, and all too often, philosophers have assumed the role of the self-appointed guardians of the natural sciences against supposed critics and sceptics, whilst themselves seeking to evade the necessity of thinking scientifically, that is, in causal and explanatory terms. Either way, so-called 'conceptual analysis' turns into cultural propaganda. The, all too predictable, result is "sophistry and illusion".[24]

---

[24] As a recent example I would cite Paul Boghossian, *Fear of Knowledge. Against Relativism and Constructivism*, Oxford, Clarendon Press, 2006. The claims may justly be called 'sophistry' because the positions that are attacked are misrepresented and the conclusions must count as 'illusions' because the claim to possess absolute knowledge is never made plausible. For justification of these criticisms see D. Bloor, "Epistemic Grace. Anti-Relativism as Theology in Disguise", *Common Knowledge*, vol. 13, 2007, 250-280.

2

Raymond Boudon

Professor of Sociology

University of Paris-Sorbonne, France

Realism: the ultimate requirement of the social and natural sciences

How did you get interested in the philosophical aspects of the social sciences?

My interest in the philosophy of social science was essentially motivated by my impression that many current products of the social sciences seemed not to aim primarily at explaining puzzling social phenomena, those we do not easily understand, in the neutral fashion that characterizes the achievements of the natural sciences. I had the feeling that many products of the contemporary social sciences are descriptive rather than explanatory. I had even the impression that some of these products inform the reader rather on the passions and opinions of the social scientist than on the real world. By contrast, I was always impressed by the numerous explanations of puzzling social facts Emile Durkheim as well as Max Weber, say, have proposed and also by their claim that the best way for a social scientist to be useful is to create solid knowledge. This does not mean that a sociologist should be indifferent to the social and political world he belongs to. A social scientist can even be a militant *besides* being a sociologist, exactly as a physicist can be a militant besides being a physicist. But there should be no confusion between the two roles. By contrast, sociology and generally the social and human sciences played often with this confusion rather than they made sure to avoid it, and not only in the last decades of the sixties as it is sometimes erroneously contended, but much earlier already. To take an example which should likely arouse few passions nowadays, Gustave Le Bon did not provide genuinely new solid explanations of collective behaviour. His pages on this topic were rather inspired by his political passions. His books have known a great success, though, because they met the similar passions of many people who also perceived the mass movements of their time as a threat to the prevailing social and political order. This ideological rather than scientific production of the social sciences was already more visible in the 19th century than the scientific one. Le Bon was much more widely known in France than his contemporary Durkheim. Max Weber had a more limited intellectual audience in Germany than ideological writers as, say, the historian Friedrich Meinecke, not only because only a few of his works were published during his life time. Earlier, Alexis de Tocqueville was much less popular in his time than much more subjective or ideological writers, as the historians François Guizot or Jules Michelet. He has never been well accepted in France, in great part probably because he had cultivated what Weber was to call later "axiological neutrality".

What happened in the 1960's and the following decades is that the scientific *ethos* was in these years overtly described as an illusion when applied to the social sciences and that many social scientists adopted this view. This relativistic philosophy is still present in many minds[1]. It explains the heterogeneous character of the social sciences and the fact that their products give rather rarely the impression of producing genuine new solid knowledge, of representing cognitive breakthroughs.

Today, sociology is less ideological than in the sixties. It has become to a large extent essentially descriptive. Descriptions of such and such corner of the social world are of course useful and they can be bright and illuminating. But the social sciences are more appealing when they succeed explaining puzzling phenomena by genuinely scientific theories, as when Adam Smith, Tocqueville and Weber, say, solved the enigma of the so-called American religious exceptionalism, when Durkheim explained why suicide rates are higher among some social categories than others, or when Weber explained why Christianity replaced the old polytheistic Roman religion.

Shortly, it is the strong impression that the contemporary social

---

[1] See my *The Poverty of Relativism*, Oxford, Bardwell, 2004, *The Origin of Values*, London/New Brunswick, Transaction, 2002 and *Renouveler la démocratie: éloge du sens commun*, Paris, Odile Jacob, 2006.

sciences have to a large extent left aside the scientific program which had given notably Tocqueville's, Weber's or Durkheim's works their strength and illuminating character that induced my feeling that an external glance on the social sciences was needed, be it called philosophy or methodology or epistemology of the social sciences. At any rate, I immediately conceived the philosophy in question as taking the form of a critical analysis of actual social scientific works: as a discipline basically devoted to disentangling the reasons as to why such and such works appear as more solid and illuminating than others. In other words, I saw it as a discipline close in its inspiration to literary or musical criticism: as directed toward definite objects, i.e. the products of the social sciences. I know that other writers write on the philosophy of the social sciences without caring much at these products. But, as far as I am concerned, I feel very sceptical on the interest of *a priori* philosophical speculations which would not take as their object of analysis the actual products of the social sciences.

Which social sciences do you consider particularly interesting or challenging from a philosophical point of view?

To me, explaining any social phenomenon means showing that it is the outcome of individual beliefs, attitudes and actions. In other words, explaining a phenomenon means: deciphering the underlying beliefs, attitudes or actions which are the ultimate causes of the phenomenon.

A main question is then: how is it possible to identify safely the causes of these actions or beliefs? The ordinary social life, the judiciary life, or successful police investigations exemplify the fact that it is currently possible to identify the causes of men's actions or beliefs, if not always at least in many cases. The most familiar episodes of social life show in other words that human motivations and reasons can in many circumstances be safely identified, in spite of the fact that are directly unobservable. If they could not, social life would be impossible. Now, the identification of human motivations and reasons is not a mysterious empathic operation. It rests rather upon the construction of a theory using the usual theory building procedures used in scientific as well as everyday life. Thus, if I observe a man cutting wood in his yard, I will reject the theory that he wants to burn it in his chimney if my observation takes place in a hot summer: the theory will be then falsified and I will have to find out alternative explanations, until

I find one which appears as more compatible with all observable data than its competitors. In other words, explaining social phenomena implies as an essential moment finding out an acceptable theory of the individual beliefs, actions or attitudes underlying the phenomenon. Such a theory should use *ordinary psychology*, i.e. the psychology used in everyday life, in judiciary courts or police investigations: the only one we recognize spontaneously as solid.

This psychology invites to find out the motivations and reasons explaining an action, belief or attitude, taking into account the context, in the broad sense of this latter word, in which the individual actions or beliefs under investigation appear. Thus, if we want to explain the behaviour of the Australian rain maker, we will have to take into account the fact that he has no competence in physics and generally no idea of science in the Western sense, and that he interprets the world as guided by spiritual forces which he wishes to seduce. Given this context, the behaviour of the rain maker can be imputed to the fact that he wants the rain to fall on his crop and that he has reasons, given his religious worldview, to believe that he can get the rain to fall through rituals the value of which are guaranteed to his eyes by a long tradition. Moreover, as he conducts his rituals in the periods of the year when rain is likely to fall, his confidence in the idea that his rituals are effective will be reinforced. The correlation is spurious indeed. But modern Westerners impregnated with scientific culture also confirm currently their causal beliefs by spurious correlations. In this analysis, I have used only 1) data on the context of the rain maker, 2) ordinary psychology. As the rain maker, modern Westerners do many things because they want to reach some goal, and they use such and such means to reach the goal because they have in mind a theory telling them that the means are efficient.

Now, ordinary psychology is by far not the one which is currently used in the social sciences. Thus, the so called Rational Choice Theory (RCT) endorses a reductionist view of human psychology[2], whereby people are supposed to be motivated exclusively by their egoistic interests: they do X if they have the impression that this should generate outcomes they see as good for them. This psychological mechanism explains obviously many human actions, but it is unable to explain for instance why I am

---

[2] See my "Beyond Rational Choice Theory", *Annual Review of Sociology*, 2003, 29, 1-21 and "Ordinary rationality: the backbone of the social sciences" (forthcoming 2008).

opposed to death penalty. As death penalty does not threaten me, why should I be concerned with it? Within the RCT framework, it is impossible to explain either why people vote, since in a general election the probability of any single vote being pivotal is practically zero, so that voting is an action without any consequence and *ipso facto* with no consequence on the interests of the voter.

Other theorists assume that human beings would always be motivated by envy, or that they would be the prey of a basic instinct of imitation, or that all their motivations or beliefs would have a sexual origin, or that they would be exclusively concerned by their class interests, or by their will to power. All these theories have generated important movements in the social and human sciences. They have given birth to intellectual sects, some of which have enjoyed a great influence; but they have never reached a universal consent.

Beside these theories, other theories maintain that psychology has nothing to do in the social sciences, since men are basically moved by social, cultural or biological forces, so that the explanation of their beliefs, attitudes or actions should not be sought on the side of their apparent reasons and motivations. Such theories can be called *holistic* since they assume that individual behaviour is the effect of *forces* emanating from the *society* or the *culture* the individual belongs to or of the *biological evolution*.

On the whole, many important movements in the social sciences have rejected the ordinary psychology currently used in social life and endorsed rather either a naturalistic-holistic view of man whereby the human being is supposed to be led by some social, cultural or biological forces, or a reductionist psychological view whereby he would be led by some basic instincts, as egoism, envy, a basic instinct of imitation, the will to power or sexual instincts. So, a basic reason of the disillusionment generated by Marxism, Freudianism, structuralism, Rational Choice Theory, culturalism or sociologism, i.e. that widespread form of sociology which tries to explain any behaviour as the effect of socialisation, lies in the fact that all these movements endorse an unrealistic view of man.

Now, science is by essence realistic. It invites to describe the world and men in particular as they are. *A contrario*, the greatness and lasting value of works as Tocqueville's or Weber's is to a large extent due to the fact that they always see the puzzling social phenomena they explore as the outcomes of individual actions, attitudes or beliefs which they analyze as caused by reasons and motivations understandable within the framework of ordinary psychology, i.e. the psychology codified or illustrated by Aristotle, the French moral writers of the 17th century, Lichtenberg or Stendhal among others.

How do you conceive the relation between the social sciences and the natural sciences?

Albert Einstein once wrote that science is a continuation of common sense. He maintained that " science is nothing more than a refinement of our everyday thinking "[3]. To the eminent scientist, there is no discontinuity between ordinary and scientific knowledge. There is no discontinuity either between the social and human sciences and the natural sciences: their basic aim, *explaining*, and their basic rules of inference are the same. The structure of explanation can in the two cases be summarized by the formula: $\{S\} \to P$. It says that explaining a phenomenon P amounts to making it the consequence of a set of mutually compatible and acceptable statements $S_1, S_2, ..., S_n$. From the moment when the phenomenon P is made a consequence of the set $S_1, S_2, ..., S_n$ of non enigmatic statements, its initial opacity is dissipated: it looses its enigmatic character and is "explained". Thus, Torricelli and Pascal have explained the puzzling phenomenon of the barometer by making it the consequence of a set of statements as: "the atmosphere has a certain weight", "the weight of the atmosphere is lower on the top than at the bottom of a mountain", etc. In the same fashion, we explain that some people believe in rain rituals because they believe that rain is commanded by spirits, and also because, as the rituals are practised in the periods of the year when rain is more likely to fall, there is a (spurious) correlation between practising rituals and rain falling, so that the rituals seem to work. And if they do not, the magician is likely to explain the failure, as the scientist, by introducing auxiliary assumptions.

In the case of the natural as of the human sciences, we have the feeling that the explanation of a phenomenon P is entirely satisfactory once its ultimate causes have been reached. Thus, the discovery that the consumption of some type of food protects against cancer invites the scientist to wonder why and to go one step further, until he finally has the impression that he has identified

---

[3] Einstein A., Physics and Reality, *The Journal of the Franklin Institute*, vol. 221, n° 3, 1936.

the ultimate causes responsible for the correlation. The approach is the same in the human and social sciences. An explanation is treated as complete once we have reached the ultimate causes of the phenomenon we want to explain, i.e. the reasons and motivations of human actions, beliefs or attitudes. Thus, Tocqueville has explained that Frenchmen at the end of the 18th century believed in Reason with capital R much more currently than Englishmen because, given the context, they had reasons to believe so that the Englishmen did not have. He explained in the same way that, again at the end of the 18th century, the British agriculture developed at a quicker path than the French agriculture because the French landowners had strong reasons to leave their land in order to occupy an official position in the State, which the British landowners did not have. This landlord absenteeism had the effect of slowing down the path of agricultural innovation in France.

As far as the identification of the ultimate causes of a phenomenon is concerned, the social sciences have even an advantage with regard to the natural sciences. Once he has identified some virus, the biologist has to go one step further, and to explain its behaviour; he has then to identify the chemical processes going on as an effect of the virus, etc., so that the ultimate causes of a phenomenon appear easily to him as being always removed one step further at each stage of his investigation. In many instances, the social scientist can by contrast reach the genuinely ultimate causes of the phenomenon he explores. To take an elementary social example: I explain the fact that pedestrians look to the right and to the left before crossing a street by the fact that they want not to be run over by a car. The explanation is complete; it gives birth to no further question. Knowing what happens in the brain or in the blood pressure of the pedestrians when this event occurs would probably not be uninteresting, but it would explore biological events parallel to the social event under investigation, rather than complement the sociological explanation or compete with it[4]. The same analysis could be made of the structure of explanations dealing with more complex phenomena, as, say, the explanation proposed by Adam Smith, Tocqueville and Max Weber of the puzzling American religious exceptionalism[5].

---

[4] See my contribution in Bronner G. (ed), *La pensée humaine en débat: Boudon, Changeux, Descombes*, Paris, Presses Universitaires de France, forthcoming 2008

[5] See my *Tocqueville for today*, Oxford, Bardwell, 2006.

This leads me to the discussion of an important issue: *materialism*. The natural sciences reached a scientific status from the moment when they were able to substitute explanations by *material* causes for explanations by *final* causes, e.g. when they were able to explain the fall of rain by material causes rather than by the will of supernatural forces. Evolution theory is the most spectacular example in this respect. Charles Darwin succeeded explaining evolutionary phenomena by making them the outcomes of material causes, while they were earlier explained by the actions of some will or wills, in other words by final causes. But the prestige of Darwinism does not justify the generalization that materialism would be a condition of any science, of social as well as natural sciences. The interest of substituting material for final causes is indubitable in the case of the natural, not of the human sciences for the simple reason that human intentions, motivations and reasons are the bricks of the human world as it is. From the moment when they aim at following a genuinely materialistic program, the human and social sciences have to get rid of all explanations including final causes and consequently to erase from the social reality the human motivations and reasons.

Strange as this may appear, this is precisely what the various influential forms of positivism (behaviourism, structuralism, economic instrumentalism,[6] culturalism, sociologism, etc.) have proposed to do. All these movements rest finally on the confusion between realism and materialism. As to the faith of their advocates in materialism, it rests on an erroneous generalization from the case of the natural to the case of the social and more broadly the human sciences. As far as the natural sciences are concerned, materialism and realism coincide with one another. In that case, being realist and being materialist is one and the same thing, since nobody was ever able to demonstrate the existence of the spiritual forces which were postulated by prepositivist thinkers. But, as far as the social sciences are concerned, ignoring human intentions and reasons and motivations is being *unrealistic*.

On the whole, the influence of positivism and the disqualification it indirectly generated of the motivations and reasons of human beings are possibly the basic factor of the sterility of many

---

[6] I.e. the theory developed by Milton Friedman according to which an economic theory should exclusively be rated by its predictive efficiency, never in consideration of the validity of the "psychological" statements it includes, since they deal with unobservable data. This theory is widely accepted among economists.

productions of the human and particularly the social sciences. For the sake of materialism they claim one should ignore human motivations and reasons or treat them as dependent variables. Thus, the confusion between realism and materialism is probably one of the causes that explains the success met by Marxism among intellectuals, since it proposed to consider that human consciousness was "false" and hence that the motivations and reasons of social actors could be ignored or treated as dependent variables. The same confusion explains also for one part the success of *Freudianism*, to which human behaviour is the effect of sexual instincts; of *culturalism* and *sociologism*, to which it should be considered as the effect of social and cultural forces; of *sociobiology*, to which it would be the outcome of biological evolution; or of *structuralism*, to which it would be the outcome of mysterious "structures".

The appeal of the materialistic postulate to social scientists is so strong that it reappears constantly even now, though in more discrete forms. Thus, some contemporary trends in the social sciences propose to explain all phenomena as the effects of the basic human instincts I mentioned earlier. Dawkins' theory of *memes* for instance, although presented in a Darwinian jargon, assumes basically that behaviour is the effect of the imitation instinct which would dominate human beings. Others propose to explain behaviour mainly as an effect of socialisation. As the postulate of materialism and the confusion between realism and materialism are deeply implanted in the mind of many social scientists, the agony of culturalism, sociologism or economic instrumentalism will likely be long.

As I have tried to show[7], the strength of the analyses of Tocqueville or Weber or even Durkheim, to limit myself to these giants of the social sciences, derives by contrast from the fact that they saw clearly that materialism and realism are distinct in the case of the social sciences: that realism does not imply materialism. They saw well that explaining any phenomenon P means explaining that, given the context, some people had reasons and motivations to act in such and such way and that their actions, once aggregated, produced the phenomenon P.

On the whole, two statements should be taken seriously. Generalizing the earlier already mentioned claim made by Albert Ein-

---

[7] *Toward a general theory of rationality: a defence of common sense*, Oxford, Bardwell, forthcoming 2008. In French: *Essais sur la théorie générale de la rationalité*, Paris, Presses Universitaires de France, 2007.

stein according to which science would be nothing more than a refinement of our everyday thinking, the *first statement* says that the procedures mobilized by the natural sciences, the social sciences and by ordinary knowledge are basically the same. Thus, the rules of inference, the insistence on realism, the disqualification of contradiction and the subsequent need to eliminate contradiction characterize the cognitive activity of the natural scientist, of the social scientist as well as of the ordinary man in his everyday life. *Second statement*: materialism is a valid postulate as far as the natural sciences are concerned, but it is invalid regarding the social sciences. Describing the natural world as animated by material causes is realistic. As to social phenomena, it is realistic to see them as produced by the reasons and motivations of men. In other words, seeing the natural world as led by non material forces was rightly perceived as superstitious from the moment when the various natural sciences were institutionalized, while, as far as the social world is concerned, the superstition lies on the side of those who propose to see men's behaviour as passively determined by biological or sociocultural forces or by psychological instincts. In that sense, many products of the social sciences, those inspired notably by structuralism, culturalism, sociologism and the other above mentioned movements, are superstitious. The biological and cognitive sciences can of course on their side explain many human phenomena which do not belong to the territory of the social sciences, as the effects on behaviour of brain lesions. The social sciences can reach the same level of scientificity when they explain the phenomena belonging to their territory provided they discard *weak concepts* –as the concepts describing occult forces– and *controversial postulates*, as the materialism postulate[8].

What is the most important contribution that philosophy has made to the social sciences?

I would place in first rank Max Weber's *Essays in the theory of science* (*Aufsätze zur Wissenschaftslehre*) among the most important works in the philosophy or methodology of the social sciences. My own work in the philosophy or methodology of the social sciences leans heavily on his in most cases very underdeveloped intuitions. He has well seen the importance of *methodological individualism*: the fact that social phenomena are ultimately the outcomes of the

---

[8] Bronner G. (ed.), *op. cit.*

understandable reasons and motivations inspiring human actions or beliefs. By contrast, holistic thinkers maintain that they would be the effect of conjectural forces emanating from the *culture*, the *society* or from *biological evolution*, so that holism treats individuals as being merely the targets of these forces. I leave aside the eclectic analysts who assume quietly the idea that human choice is rational while human behaviour would be irrational: that people choose their means on the basis of reasons, but follow a goal or endorse a representational or normative belief under the effect of some social, cultural or biological forces. To Weber by contrast, one can *understand* why an individual assumes some representational or normative belief or some goal as well as why he chooses some means. Otherwise, he would not have attempted to show in his *Essays in the sociology of religion* (*Aufsätze zur Religionssoziologie*) that people in such and such context had understandable reasons to believe what they believed. Weber has also had the intuition that ordinary psychology is the basic tool to be used in the understanding of human behaviour. For this reason, he was very critical of psychoanalysis and Marxism. He had moreover the feeling that rationality could not be reduced to instrumental rationality and created consequently the concept of axiological rationality. True: he used the concept only twice in his work and never explained what he had exactly in mind when using it. The concept has been almost forgotten since. It has sometimes been treated with condescendence even by writers sympathetic to Weber.

I have instead tried to show for my part that, once properly elaborated, this concept contains in a nutshell a powerful criticism of the shortcomings of the dominant theories of rationality, as the Rational Choice Theory[9]. Can we notably accept the idea that human beings would be rational in the choice of their means and socially determined in the choice of their beliefs, goals, norms or values? A proper elaboration of Weber's rough intuitions on rationality can provide a general theory of rationality which overcomes the shortcomings of RCT without falling into the trap of eclecticism: choice explained *rationally* as the effect of understandable

---

[9] See my "Are we doomed to see the *homo sociologicus* as a rational or as an irrational idiot?". In Elster J., Gjelsvik O., Hylland A., Moene K., *Understanding Choice, Explaining Behaviour, Essays in the Honour of O.-J. Skog*, Oslo, Unipub Forlag, Oslo Academic Press, 2006, p.25-42 and *Quelle théorie du comportement pour les sciences sociales ?*, Conférence Eugène Fleischmann III, Paris, Société d'ethnologie, 2004.

reasons, behaviour explained *irrationally* by the action of social or cultural forces. In my elaboration, I started from the postulate that we find a norm or a value right or wrong or a theory true or false once we have reasons to do so. Beside *instrumental rationality*, we have in other words to take into account the form of rationality which can be called *cognitive*. Max Weber's instrumental rationality says that we may endorse a norm or a value because it generates outcomes we perceive as positive; *axiological rationality* that we can also endorse a norm or a value because it is grounded on principles we have reasons to see as positive. Moreover, we perceive these reasons as valid only if we have the impression that others would also see them as valid, exactly as we feel that a mathematical statement or an empirical theory is true only if we have the impression others would approve our reasons of seeing them as valid. I recognize that I went much farther than Weber himself, when I proposed to see *axiological rationality* as a particular declination of *cognitive rationality*. But I do not care whether Weber himself would have endorsed my elaboration of his intuitions, since such a question has no answer and hence little interest, to me at least.

Beside Weber, another writer impressed me, namely Karl Popper, not so much by his celebrated theory of falsification, which seems to me difficult to accept at least literally, than through his little known article on "the myth of the framework"[10]. The point of this article is that the social sciences are wrong when they think they have explained some behaviour by saying that a man behaved in such and such ways because he had in mind some mental "framework". The criticism goes very far. It deprives of any genuinely explanatory value most dispositional concepts, as "social representations", "cognitive biases", "*habitus*", "mental frames" or "cultural frameworks" and proposes instead to see the so-called frames or frameworks, not as indelibly inscribed in the mind of human beings, but rather as *conjectures* which are not questioned by them only as long as they seem not to be refuted by the real world or by the strength of some arguments.

Max Weber has been in advance aware of the importance of Popper's point. This can be illustrated for instance by the way he explained why the Roman civil servants left easily the old poly-

---

[10] Popper K., "The Myth of the Framework", in: Freeman (E.), (ed.), *The Abdication of Philosophy : Philosophy and the Public Good,* La Salle, Ill., Open Court, 1976, p. 23-48.

theistic Roman religion for the monotheistic cults introduced on the Roman market of religious ideas by actors coming from the oriental part of the Roman Empire. The civil servants, Weber explains, had the impression that the new cults which saw the world as dominated by a single powerful will provided a much better symbolic representation of the social order illustrated by the Empire they served than the polytheistic religion. The Roman peasants by contrast remained attached to the old polytheistic religion because the monotheistic idea of a world regulated by a single will appeared to them as incompatible with the capricious character of the events important to them, notably the meteorological events. So, the civil servants had reasons to prefer the new religious beliefs to the old ones while the peasants had reasons to reject them, exactly as a scientist has or has not reasons to prefer a theory to an alternative one.

The idea that *frames* or *frameworks* would be indelibly inscribed in human minds under the influence of social and cultural forces is to Popper a mere *myth*. This powerful idea invites to nothing less than a global revision of the social sciences. It rejects the simplistic idea that people would uncritically and passively accept some ideas, norms or values under the effect of various forces: an idea common to Rational Choice Theorists, Marxists, Freudians, structuralists and culturalists.

I appreciate Mario Bunge's work for his impressive effort to incite social scientists to be more scientific and less ideological[11]. I agree with him when he states that the social sciences can be and have been in many occasions as scientific as the natural sciences. But I disagree with him when he wants the social sciences to be materialistic, as the natural sciences. I disagree for the reason already mentioned that materialism is realistic as far as the natural sciences are concerned, but unrealistic in the case of the social and generally the human sciences.

Which topics in the philosophy of social science will, and which should, receive more attention than in the past?

Many topics seem to me important and likely to give birth to interesting research in the philosophy – or methodology – of the

---

[11] Bunge M., *The Sociology-Philosophy Connection*, Londres/New Brunswick (USA), Transaction, 1999; *Chasing Reality. Strife over Realism*, Toronto, The University of Toronto Press, 2006.

social sciences. I will just mention a few ones. I do not come back to the crucial question I raised earlier: which theory of behaviour for the social sciences?

Another topic would deal with the forms and uses of the social sciences. I leave aside the type of social sciences inspired by the scientific *ethos*. The social sciences can also have the interest of drawing the attention on unnoticed socially important issues. This is the function of the so-called *critical sociology*. Critical sociology is socially useful. But it becomes easily counterproductive when it presents itself as a substitute to scientific sociology[12]. Another type is cameral sociology: it aims less at explaining puzzling phenomena than at producing useful data for the various decision makers. Another type is aesthetical sociology. It aims at arousing emotions rather than at explaining social phenomena. The interaction between these forms of sociology would be worth being explored. Thus, the underexploitation of survey data is the effect of the fact that they are produced for cameral purposes. Take the example of political surveys: the political men or parties for whom they are produced are above all interested in *how many* voters and *which* voters approve them. Explaining *why* they do or do not approve them implies more costly surveys, introducing more variables and a greater sophistication of the analysis. Of course these types are ideal.

Another interesting topic is the following. In the natural sciences, the word *theory* has the clear meaning I referred to earlier: it describes a set of statements compatible with one another and individually acceptable thanks to which a phenomenon is explained. In sociology, the word *theory* means by contrast many things. It is sometimes used to qualify a concept or a set of concepts, or as a synonym of the concept of paradigm. Sometimes, it qualifies sketchy descriptions, distinctions or sweeping generalizations, as when modern societies are depicted as *risk societies* or as *consumption societies*. Often, descriptive concepts, as *social capital*, *frame*, *framework*, *habitus* or *social representation* are perceived as containing a genuine theory, while they are essentially useful descriptive labels: a costly confusion, as indicated by Popper's previously mentioned article on "the myth of the framework". Describing this polysemy and disentangling the meaning of the polysemy of the notion of theory in the social sciences would be

---

[12] See my "Sociology that really matters", 1st Annual Lecture, European Academy of Sociology, *European Sociological Review*, vol 18, no. 3, 1-8.

an interesting project.

Another meaning of the word theory is particularly interesting: it is sometimes taken as a synonym of *interpretation*. Thus, the Marxist interpretation of the French Revolution which proposes to see it as an episode of the class struggle which would pervade the history of mankind is easily christened *theory*. So, an important issue would deal with the difference between the notions of *interpretation* and *explanation*. The explanation of a phenomenon aims at being unique and true. By contrast, several interpretations of a given phenomenon can easily coexist, though some interpretations can be more acceptable than others. On which criteria can statements as "the former interpretation is more acceptable than the latter" be grounded? Are there types of phenomena which have to be interpreted rather than explained? If yes, which ones?

It would also be fruitful to explain why positivism has been and remains so influential in the human and particularly in the social sciences and generally, why many movements, as structuralism or today constructivism, were so influential although their contribution to knowledge was more or less rapidly held as highly questionable. Clarifying these questions is also likely to make the social sciences more self-conscious.

On the whole, the interesting questions which can be raised are so numerous that the philosophy or methodology of the social sciences should be more widespread than it is presently, where it suffers from a kind of marginal status. Even worse: under the influence of the anomic division of intellectual labour Durkheim has already stigmatized, it is held as a special field beside the sociology of organizations, the sociology of work or the sociology of the Middle East, say, while it should be a common denominator of all sociological fields, exactly as mathematics is a common denominator of all fields in physics. Taking this diagnosis seriously would imply a drastic reformulation of many social science curricula. This might be the price for the consolidation of this crucial sector of knowledge.

3

Mario Bunge

Professor of Philosophy
McGill University, Canada

How did you get interested in the philosophical aspects of the social sciences?

My interest in the philosophy of science arose back in 1936 in my native Argentina. The high-school curriculum included Logic and Philosophy of Science. Logic was limited to syllogistics, and the philosophy of science was restricted to safe and outdated problems such as the classification of the sciences. Still, this sufficed to spark off my curiosity, indeed my passion, for the subject. It even helped me choose a career. The story is as follows.

For a while I hesitated between psychology, philosophy, and physics. My interest in psychology was awoken by psychoanalysis, which was then taken for granted in my circle. Freud's books were sold for dimes at subway kiosks, largely because it was mistakenly believed that he was an expert on sex, which of course was of outmost importance for a teenager. At the time there was no one to challenge psychoanalysis in Argentina. Sadly, this situation has remained unchanged after seven decades. In my country of origin I am better known as an excentric and fierce critic of psychoanalysis than as a philosopher. However, let me return to the story of my career choice.

I was also strongly attracted to philosophy, which I read widely in several languages. I was stunned by the pre-Socratics, bored by Plato, awed from afar by Aristotle, disappointed by Descartes and Leibniz, marvelled by the Encyclopaedists, intrigued by Hegel, and enthusiastic about Engels and Russell–odd bedfellows that they were. I was lucky to stumble on Russell's *Problems of Philosophy*, which contains a clear discussion of Pavlov's physiology and Watson's behaviorism, of which I became an instant convert.

My next decisive reading was Reuben Osborne's *Freud and Marx*, which had recently been circulated by the Left Book Club.

I found it inconsistent with both Marxism and the tiny fragment of psychology I had just learned, and said so at book length. (Fortunately the typescript got lost along with two novels and a play.) That exercise convinced me that psychology was not for me. (My interest in the subject was rekindled more than two decades later, when meeting mathematical psychology at the University of Pennsylvania, where I served as a visiting professor. At the time I still believed that if it's mathematical, it must be scientific–an illusion I criticized in 1967 in my *Scientific Research*.)

Disillusioned by psychology, I gave university philosophy a try: I started auditing the freshman philosophy lectures at the Universidad de Buenos Aires. But I gave up after one week, for I found the professors outdated and hostile to science. (Some of them admired the intuitionist Henri Bergson, and others were followers of the neo-Hegelian and fascist Giovanni Gentile.) I did not return to the same faculty until twenty years later, as the professor of philosophy of science.

At about the same time I became fascinated by astrophysics through the popular books by Sir Arthur Eddington and Sir James Jeans. I loved their books but rejected Eddington's Kantian subjectivism and Jeans's Platonism and religiosity. I could not believe that physicists only discover what had been in their minds all the time, as Eddington claimed, because I had performed a few experiments; nor could I believe that the universe is a vast collection of mathematical formulas, as Jeans held, because I did not meet any mathematical objects in my parents's large garden cum orchard. At the same time, I realized that I lacked the knowledge required to criticize those outlandish philosophies of physics. This is why I decided to study physics, which I did. I worked as an instructor in theoretical physics for five years, at the end of which I was sacked for refusing to join the Peronist Party. I eventually got my PhD in physics under Guido Beck–an Austrian exile–, and was a professor of theoretical physics, on and off, in Argentina and in the US, between 1956 and 1965. My first physics paper appeared in 1945, and my latest in 2003.

In short, my early love of philosophy led me to study physics, which in turn led my back to philosophy. I went on studying philosophy on my own, and wrote a few philosophical papers, the first of which was published when I turned 20. In 1944 I founded the philosophical journal *Minerva*, that lasted only six issues. I also organized a philosophical circle of amateurs like myself, which met once or twice a month. Although we were all interested in politics, and even victims of it, none of us knew any social science, much less philosophy of social science. We only discussed problems in the philosopy of mathematics, physics, biology, psychology, and technology.

Initially I embraced the positivist (or Copenhagen) interpretation of quantum mechanics, according to which the theory deals with object-apparatus-subject systems rather than with physical things in themselves. But upon reflecting on free electrons it suddenly dawned on me that the interpretation in question was false. I then resolved to construct a realist interpretation, a task I only completed in 1967, with my *Foundations of Physics*.

My interest in the philosophy of social science came rather late. I had always been interested in social problems and in politics–who is not in a backward country? But I had never read any serious works in basic social science, except for Albert Mathiez's history of the French revolution and a few things in anthropology. In 1957, when I took the chair in philosophy of science at the Universidad de Buenos Aires, I realized that I was not qualified to discuss any philosophy of social science. Since I did not believe in doing philosophy of X without knowing some X, I asked Gino Germani, the founder and director of the Department of Sociology at the same school, to recommend me publications in his field. Gino recommend Parson's book on systems of actions. I disliked this work because of its remoteness from topical social issues, imprecision, pomposity, and holism. So, for a while I desisted from learning sociology. Luckily, a few years later I stumbled on Robert Merton's *Social Theory and Social Structure*, which I admired instantly. (We became friends three decades later.)

However, to a cheeky Argentinian the best way to learn a subject is to teach it and write a paper on it. So, in 1967, soon after arriving in McGill University, my final academic home, I offered a seminar on mathematical sociology using the book that James S. Coleman had recently published on the subject. I went to see the chairman of the Department of Sociology and asked him to send me some students. His answer was: "Mathematical sociology? Never heard. What's that?" Still, I did teach the seminar to half a dozen non-sociology students.

At about that time I wrote a paper on something about which I had a personal experience. It consisted of four elementary mathematical models of human migration. Later, while studying Piélou's book on mathematical ecology, I got some ideas on social structure, which I presented in a longuish paper published in 1974.

Shortly therafter I drew the attention of Máximo García Sucre, a Venezuelan quantum chemist who was auditing my metaphysics course, to this old problem: What keeps social systems together despite the different, often conflicting, interests of their components? Building on my study of social structure we wrote a paper on participation and cohesion, that got published in *Quality and Quantity*.

All that amateur work in sociology sparked off my interest in the philosophy of social science. As in previous cases, I was putting into practice my motto:*Primum cognoscere, deinde philosophari*. I also put into practice the idea that mathematics is the pass key for openening the gates to all reserch fields. But I soon found out that this holds only for highly formalized domains, not for fields overgrown with wild weed—undigested empirical data.

However, I still had to overcome a tall hurdle: my awe of economics. I had been impressed by Leontief's input-output matrices in macroeconomics, as well as by the wealth of symbols in the books by Samuelson and others. At the same time I realized, just by reading the newspapers, that the economic theories of the day did not match economic reality. For example, the persistence of unemployment seemed to falsify the dogma of general equilibrium; and stagflation (inflation with economic stagnation) falsified Milton Friedman's theory. Suspecting that economics was only a paper tiger, one day I took the plunge and started reading economics journals at a furious rate. I soon confirmed my suspicion: that standard economic theory did not match economic reality.

Worse, the reading of John Maynard Keynes's admirable *General Theory*, as well as some of the criticisms of Joan Robinson, Oskar Morgenstern, and John Blatt (whom I respected for his work in nuclear physics) showed that some of the central concepts of the theory, those of capital, value, and utility, were not mathematically well defined. So, standard economic theory was neither conceptually clean nor empirically confirmed. I said so in my book *Economía y filosofía* (1982).

I repeated this contention to a packed audience at the first Spanish congress of philosophy of science at Oviedo in 1982. This provoked an irate reply from two local professors of economic theory. One of them attacked the portable blackboard with such fury, that it dissembled and fell down with a loud noise. At this I exclaimed: "The bankruptcy of standard economic theory!" Laughter and applause. The professor then asked plantively: "But then, what shall we teach?" He was unwittingly proving my point, that standard economic theory had become an academic industry, using mathematical and peudomathematical formulas more to persuade and intimidate than to explore reality.

My first essay in social ontology occupies a long chapter of Volume 4, *A World of Systems*, of my *Treatise on Basic Philosophy* (1979). And my first comprehensive analysis of the social sciences appeared in Volume 7, Part II, of the same work (1985). There I examined some methodological problems in anthropology, linguistics, sociology, economics, politology, and history. I expanded on this work in my subsequent books *Finding Philosophy in Social Science* (1996), *Social Science Under Debate* (1998), and *The Sociology-Philosophy Connection* (1999). These books were reviewed in major social science journals but ignored by all the major philosophy journals–as were all my other books.

Which social sciences do you consider particularly interesting or challenging from a philosophical point of view?

I regard all of them equally interesting, challenging, and problematic. However, at the present time I would single out the newly emerging intersciences, such as socioeconomics, political sociology, and ecological economics. In *Emergence and Convergence* (2003) I argue that the merger of previously separate disciplines, though less frequent than their splitting into narrower fields, is scientifically and philosophically important because it erases artificial barriers and helps connect different levels or aspects of one and the same piece of the real world.

How do you conceive the relation between the social sciences and the natural sciences?

In principle there are four possible relations between the natural sciences (N) and the social sciences (S):

1. *Hermeneutics*: S and N are mutually disjoint. Counterexamples: all of the biosocial sciences, such as social psychogy, neurolinguistics, demography, epidemiology, medical sociology, and human geography.

2. *Naturalism* (e.g., sociobiology and evolutionary psychology): S is included in N. Counterexamples: all of the social inventions and conventions that erect obstacles to the fulfilling of biological needs and wants–from extravagant kinship rules to property rights

to wasteful ceremonies. (Incidentally, many philosophers confuse naturalism with scientism.)

3. *Sociologistic reductionism* (e.g., the Strong Programme in the sociology of science). This is the thesis that all of the sciences, even mathematics, are "through and through social" just because they are collective undertakings. All the evidence points against sociologism. Indeed, the data and hypotheses of biology refer to organisms regardless of social circumstances, and they are tested by socially neutral methods. (Incidentally, any proof of this assertion requires a theory of reference, such as the one I supplied in the first volume of my *Treatise*.) Of course biology (just like physics) may be used or misused to bolster or undermine the philosophical kernel of any ideology. But this is another matter.

4. *My own view is this: N and S partially overlap and interact*: With regard to the N-S distinction there are three kinds of discipline: purely natural, such as physics and molecular biology; purely social, such as sociology; and biosocial, such as the interdisciplines listed above when criticizing hermeneutics. However, since the social sciences study social facts, and these involve human animals, every social fact involves biological processes that the social scientist may ordinarily overlook but cannot deny. For example, the criminologist who studies small-scale theft (in contradistinction to large scale theft such as colonialism) knows that it may be related to such physiological processes as hunger and drug addiction, or perhaps to processes in the underdeveloped prefrontal cortex of an adolescent who has not yet internalized the social conventions about private property. So, occasionally the social sciences have to use some natural sciences. Shorter: All social facts involve natural facts, but to a first approximation they may be studied the way the hermeneuticists and Emile Durkheim wanted, namely as purely social (or cultural).

What is the most important contribution that philosophy has made to the social sciences?

I believe that philosophers have made three important contributions to the social sciences. The first is Aristotle's thesis that politics and morals are linked, if not *de facto* at least normatively: that politics should be the strong arm of ethics. This link has been tacitly ignored by the utilitarians, pragmatists, and Marxist-Leninists. But it should be restored if we want politics to serve the common good rather than private interests—Halliburton's, for example. True, we cannot accept Aristotle's ethics and politics, for both of them were elitist rather than universalist. But we should admit that every political action should be judged not only practically, by its goal and means, but also morally. For example, when analyzing terrorism we should not only condemn it in all its forms, but we should condemn state terrorism even more strongly than group-sponsored terrorism because states, unlike unlawful militias, are expected to abide by international law rather than violating it serially. This condemnation is moral because military aggressions of all kinds are criminal acts (as defined by the UN), and crime is immoral because it is antisocial.

The second important contribution of philosophy to the social sciences is due to Kant, the neo-Kantians, and hermeneuticists. It is the thesis that the social studies cannot be scientific in the same sense as the natural ones, because they deal with spiritual *(geistige)* facts—hence the name *Geisteswissenschaften*. And, since the subject of the social studies is reputed not to be out there, like stars and frogs, their method must be different too: it must discover the "meaning" (meaning intention or goal) of the actors in question.

I submit that the idealist contribution has been definitely negative, because it has retarded the development of the social studies. Now any literary critic feels entitled to pass judgment on social science—criticizing, for example, the use of the so-called "numerical paradigm", the construction of mathematical models, and the search for objective indicators of social processes. (Raymond Boudon tells me that some smart German librarians class that stuff under *Soziale Belletristik*.) Unsurprisingly, hemeneutics has been scientifically barren. Worse, it has slowed down the pace of social science by espousing radical ontological individualism: it denies the objective existence of social systems such as business firms, schools, and states, that have (systemic or emergent) properties of their own.

However, the damage caused by that antiscientific school has been limited because most social scientist do not read philosophy. In fact, the entire *Verstehen* literature is more programmatic than substantive. Even Max Weber, the first (and last) important social scientist to praise the hermeneutic approach, did not practice it. For example, in 1916 he pressed for the continuation of the First World War because he was a rabid nationalist, not because he delved into the intentions of the butchers on either side.

Finally, according to the Marxist-Leninist catechism, Marx and Engels made two enduring philosophical contributions: dialectical materialism and historical materialism. I submit that the former, an ontology or metaphysics, is irremediably flawed because dialectics is at best false, and at worst imprecise to the point of meaninglessness—as I argued in my *Scientific Materialism* (1981). By contrast, historical materialism has a true important kernel: the thesis that people are ordinarily moved by material interests, whence the social scientist should always ask, like the criminal detective: Who benefits? For instance, did the Hellenes attack Troy because Paris eloped with Helen, or because they coveted the trade route to Cyprus, the copper island? And do the US governments support whatever Israel does just out of love for the Jewish people, or because Israel is defending the gates to the greatest oil deposits in the world?

I believe that historiography has followed the path blazed by Marx. Witness the most productive and influential of all the historiographic schools of the twentieth century, namely that of the *Annales* (Marc Bloch, Lucien Febvre, Fernand Braudel, Pierre Vidal, and many others). This school is definitely materialist, and it goes much farther than economism. In fact, it also looks at the physical environment, population changes, and food preferences. Something similar can be said of serious anthropologists: they start by asking how people earn their living, and only then investigate their ceremoniess. Ditto social archeology: it tries to figure out how ancient peoples eked out their sustenance and how they organized themselves.

True, sociology, economics and politology have not benefited in the same way from historical materialism. This may be due not only to an idealist bias on the part of the scholars concerned, but also to the fact that few Marxist have made any original contributions to those disciplines. In particular, no one has given solid evidence for the thesis that all change, in particular all social change, originates in conflict. Ironically, only orthodox economists and rational-choice theorists–all of them anti-Marxists–believe that every social fact is "ultimately" a market event: that there is a marriage market; that schools are markets where pupils trade homework for grades; that organized religions compete with one another in the religious market–and so on. Such extreme economism would have repelled even Marx, who believed in the power of ideals and in the intrinsic value of art and science.

To conclude this section: Philosophy cannot help influence the social sciences because these, just like the natural sciences, presuppose general philosophical theses about the nature of reality and the ways to spy on it. But whether the influence will be constructive or destructive, profound or shallow, depends entirely on the kind of philosophy.

Which topics in the philosophy of social science will, and which should, receive more attention than in the past?

Not having prophetic skills, I have no idea where the philosophy of social science is going. On the other hand, I have a few proposals for research projects:

1. What is the present status of experimental economics? Given that most if not all experiments in economics have used exclusively undergraduate students in developed countries, what are their universality credentials? Is it likely that different results would be found in other social groups and in less developed regions, such as Haiti and Bangladesh? Shorter: Examine some of the most important experiments in economics (such as those of the Zürich school) to check whether the subjects's responses depend critically upon their economic level and social status–as some early experiments in social psychology suggested.

2. Social scientists, just like ordinary folks, face inverse problems all the time. That is, they have access to observable behavior, not to the underlying mechanisms. The latter must be guessed, and these guesses must be testable to qualify as as scientific. And yet philosophers of science have consistently ignored the very existence of inverse problems.

3. Empirical testing involves the construction and use of indicators, i.e., bridges between unobservable variables and observable ones. (Example: The demand for artificial fertilizers is a reliable and easily accessible indicator of gross agricultural production in the US because nearly all American farmers use them to increase crops, and the fertilizer industry, unlike agriculture, is highly concentrated.) Yet few philosophers of social science have dealt with social indicators. Even fewer have noted that empirical indicators are bound to be ambiguous because they are not supported by any hypotheses showing the posssible mechanism whereby

the changes in the observed variable are caused by changes in the hidden variable. Are there necessary and sufficient conditions for a social indicator to be unambiguous?

4. Most of the key concepts of social science–such as those of utility, class, equality, justice, market, capital, risk, liberty, and democracy–are rather vague. How can one hope to exactify them? By linguistic analysis, logical analysis, mathematization, inclusion in a hypothetico-deductive system, or all of the above?

5. Which if any are the philosophical presuppositions of the various social sciences?

6. Most of the data and hypotheses in social science are at best approximate, that is, only partially true. Yet there is no generally accepted theory of partial truth. Worse, the popular theory that identifies degree of truth with probability (or else with improbability) is false, if only because the concept of truth is epistemological or semantical, whereas that of chance is ontological. Can anyone do better than that?

7. The concept of democracy is still rather young and not as generally accepted as one might wish. For example, the statesmen who have no scruples in ordering military aggressions may believe in the need to defend democracy at home or even to export it–by undemocratic means. Are there any universal moral principles to criticize such double standards and justify democracy in any way other than by its success?

8. The so-called policy sciences must have moral presuppositions since they are engaged in planning changes bound to affect many human lives. Which are those presuppositions?

9. Ideology is generally regarded as the opposite of science. Yet at the same time most people admit that they uphold some ideology or other, and even that they can offer good reasons for preferring their own. Is a scientific ideology conceivable? If so, what would it look like?

10. From the very beginning, the philosophy of the social studies has been plagued by the individualism/holism dualism. Yet most practitioners of social science have never been paralyzed by such dualism: they have studied individual behavior to understand the formation and dissolution of social

wholes, and they have studied the latter to understand individual behavior. Which are the concepts and hypotheses that bridge the micro to the macro?

11. Gerald Debreu earned the Bank of Sweden ("Nobel") prize in economics for constructing his axiomatic theory of general equilbrium. Was he right in claiming that the mathematization of a theory renders it impregnable to empirical tests? How is a mathematical model of a piece of reality related to the latter? Through semantic assumptions, indicators, or both?

This list suggests that the philosophy of the social sciences has a long way to go before it reaches the level of sophistication attained by the philosohies of physics and biology. What can be done to raise the level of discourse? You tell me.

4
Nancy Cartwright

Professor of Philosophy

London School of Economics and Political Science, UK and University of California at San Diego, USA

How did you get interested in the philosophical aspects of the social sciences?

I've always been interested in social science, primarily as an aid to social policy. As a child I was taught to revere Franklin Delano Roosevelt. Then I grew up with the marches on Selma, JFK, President Johnson's visions of the Great Society, Head Start and the like. We thought, like Otto Neurath, that society could be improved and that social science would help show us how. As a graduate student I was deeply influenced by Kathryn Pyne Addelson, who left Chicago Circle to work with the distinguished sociologist, Howie Becker, famous for his participant-observer studies on deviance.

I didn't work on philosophy of social science early in my career because it is so difficult to do well. Philosophy of physics is easy if you are good at maths. Physics itself is compact and uncontroversial in contrast with the social sciences, and the philosophy of physics is already highly structured. It's not hard to find a well-defined problem where you can be pretty sure if you work hard on it you'll have something interesting to say, a reasonable 'contribution to knowledge', in a reasonable time. Not so in general with the philosophy of the social sciences.

So I never ventured there till I was appointed at LSE. Then I thought it was time to try to tackle some problems in the philosophy of social science. Inspired by Mary Morgan and the exciting work on causation by the early econometricians she introduced me to, I started with issues on causation in economics, where my background in probabilistic causality and in maths gave me a bit of a push-off.

LSE is naturally a fantastic place to be located for philosophy of the social sciences. Not only are the social scientists first rate, but it is small enough to get to know many of them, across different disciplines, and there is still a widespread commitment that in the end, albeit maybe indirectly, what we learn in social science will be useful to solving social problems.

At LSE I have discovered that it is good in many respects to be a dilettante, as I am in the subject. Like almost all good scientists, LSE social scientists are highly focussed and extremely expert at what they do, and they know how to do it well. But concomitantly they have little understanding, and too often, little respect, for the methods of the other disciplines. I admiringly spoke of Mary Douglas at lunch one day; I thought my colleague from Finance was going to stand up and hit me. (When I told of this at home, my 8-year-old daughter was worried: 'But mama, he couldn't have hit you in the Senior Dining Room, could he?') The trouble with this Balkanization is well-known: Most social problems do not fall into one discipline or another and in general cannot even be described properly in the regimented languages appropriate to science. So there's a real place for knowing a little across the disciplines and trying to figure out how to use the tools they together provide to tackle social problems.

Which social sciences do you consider particularly interesting or challenging from a philosophical point of view?

I am particularly interested in the use of social science to help solve social problems. For this I take it that all the social sciences are necessary and none is more fundamental, more universal nor uniformly more accurate than any other. Even together they are not sufficient. Instead their various results and methods must be bent, extended, made more concrete and artfully combined with varieties of 'non-scientific' knowledge before we can expect to produce good, predictively successful accounts of real social phenomena, and even then these may well be rare, limited in scope and in need of a great deal of ongoing tinkering (a bit like my old Rover).

I have a great deal of confidence in methods within most of the social sciences. The methods are well-honed and seriously worked over. In the hands of skilled practitioners they can give good answers to special scientific questions. (There is naturally a feedback loop where we try to develop methods to help with the questions we want to ask, but that is not so easy to do.) So the kinds of questions we answer in the social sciences are constrained by the methods we have to answer them as well as by the special demands for reliability, precision, accuracy and lack of ambiguity that characterize the sciences, whether natural or social.

The trouble with policy is that policy questions aren't generally the kinds of questions our well-honed methods are good at answering. Indeed, the questions generally are not even properly posed in the tight kind of language required for science. So the use of social science to help with social problems is a far more complex, variegated and ultimately uncertain matter than what we are used to within the social or natural sciences themselves. This is where I think special effort is needed right now and it is a place where philosophy should be especially helpful. We need to develop far better accounts of how to combine what we know, from 'pure' science and elsewhere, to make better predictions about the characteristics we care about in real social systems embedded in their natural settings.

How do you conceive the relation between the social sciences and the natural sciences?

I am a follower of Otto Neurath on the relations of the sciences. There is no pyramid, no hierarchy, no natural order among them. They are like balloons whose boundaries can be stretched or molded and that can get tied together again and again in different bundles in different ways to address different problems.

Unlike Neurath, however, and more like John Dupré, I am willing to speculate that the same is true in nature itself. There are no sets of more basic properties or laws that fix everything at a higher level. Nature looks to have millions of different features that interact in a large number of different ways. Sometimes they can be regimented into reasonably closed systems where a certain set we single out for attention interact only with each other and produce predictable results, even perhaps on features outside the set. The venerable example here is the grand mechanical clocks of the Mechanical Philosophy, where the inner workings cause mechanical figures to appear and disappear at set times, clanging cymbals or beating drums. But these are special circumstances not the recipe for all of nature.

The important divide for me is between the sciences – both natural and social – on the one hand and the practical uses to which they might be put on the other, what might be called 'applied' science if the label did not make it look too easy. Is predictability less in the social sciences? Probably. But that may well be because the social sciences, as Weber urged, are torn between trying to be scientific on the one hand and trying to understand the phenomena we care about on the other. But as Weber pointed out, the phenomena we care about may not fit within the strictures of science and they may well not, even in principle, be reducible to (or 'supervene on') features that do.

The social sciences may not predict as well but that can be because they do not abandon their original topics and substitute new more orderly concepts and they are less able to build and protect the systems they know how to (like batteries in their casings) but are expected to fix up the systems they are given. The Nobel prize-winning econometrician Trygve Haavelmo remarked on this: No-one asks a physicist to forecast the trajectory of an avalanche but we economists are expected to forecast the trajectory of the economy.

What is the most important contribution that philosophy has made to the social sciences?

I'm sorry. I don't have any very good ideas here.

Which topics in the philosophy of social science will, and which should, receive more attention than in the past?

Right now we need more effective accounts of how to put the knowledge we have to use. Concomitantly we need more interdisciplinary cooperation. There has for a long time been a great deal of lip service to interdisciplinarity. But interdisciplinary work is extremely costly in time, money and effort, and there is little genuine support for it. Often advocates hope to get away with sponsoring a few regular interdisciplinary seminars. But what it generally takes is long-term, regular, dedicated interaction and that is bound to detract from the in-discipline researches themselves. So scholars need real support and real motivation to undertake interdisciplinary work and to do the job well. Wartime has often provided these – cf. the MIT efforts on the radar that Peter Galison describes or Neurath's studies of war economies. But many peacetime problems are just as severe and just as intractable. We need war-time levels of support for interdisciplinary work to tackle them.

5

Margaret Gilbert

Abraham I. Melden Chair in Moral Philosophy
University of California, Irvine, USA

Which social sciences do you consider particularly interesting or challenging from a philosophical point of view?

Those disciplines referred to as the social sciences – including sociology, social anthropology, and economics – are relatively diverse in terms both of their concerns and their methods. It is therefore not surprising that the philosophy of the social sciences has included a wide range of questions. Some are specific to a particular social science; others are more general, such as the question given the nature of their subject matter: are there major differences in principle between the social and more broadly speaking human sciences and other scientific disciplines such as physics?

The question that has most interested me is a related general question. It is this: what are the contours of the *social* domain? This question much concerned the acknowledged founders of sociology, Max Weber and Emile Durkheim, along with Georg Simmel, another classic figure in that discipline. That these classic authors gave divergent answers is well known. Irrespective of social science, or the philosophy of social science, this question and those that come out of it constitute, I would argue, an important part of philosophy itself, a part that might be labeled *the philosophy of social phenomena*.

How did you get interested in this question?

My interest in this question developed gradually as the result of a number of prompts, some quite fortuitous. A review of some of these may serve to introduce the question, others to which it leads, and others to which it is pertinent.

Not long after I had taken up my first teaching post in philosophy a sociologist friend introduced me to Weber's concept of social action. There is a social action in Weber's sense when, roughly, one person does something with another person in mind. For instance, a man turns a corner in order to avoid having to greet an acquaintance he has just sighted. According to Weber, social actions are the central subject matter of sociology. I reacted critically to this idea and my friend and I had quite an argument about it.

I do not recall what my criticisms were at that time, but one might argue that a single social action in Weber's sense is far from a paradigmatic social phenomenon. The central subject matter of a discipline concerned with such phenomena must surely be characterizable in other terms.

Perhaps some kind of complex of Weberian social actions would do better in this regard than a single social action. In this connection I became interested in David Lewis's book *Convention: A Philosophical Study* to which I was serendipitously introduced by another friend when we met on a plane.

Inspired by a proposal from the economist Thomas Schelling, this book offered an account of the everyday concept of a social convention that was couched in terms of the mathematical theory of games. I was impressed by the clarity and the intricacy of the discussion. I was even more impressed by Lewis's discussion of the phenomenon he referred to as "common knowledge". Roughly, when something is common knowledge between two or more people each knows it, each knows it, and so on. The concept of common knowledge was clearly an important addition to the conceptual apparatus in terms of which human social life was to be understood.

There was at the time a general tendency of philosophers to say that what they were discussing – language, say, or morality – was a social phenomenon, without saying what it was to be a social phenomenon. In his famous book *The Idea of a Social Science*, Peter Winch had said that it was important to examine this question, but he did not seem to have done so. Already stimulated to think about it by my encounters with Weber and Lewis, I decided seriously to pursue it.

My initial conclusion was that sociality came in degrees. I took seriously Weber's suggestion that a social action in his sense had something social about it. If that were the case, however, a Lewisian convention – a complex structure of mutual expectations and preferences of which there was common knowledge in the population in question – seemed to be far more social. Indeed, such conventions appeared to be among the pre-eminently social phenomena.

In 1978 I incorporated this position in an Oxford University doctoral thesis with the title *On Social Facts*. Shortly afterwards, I received a contract from the publishers Routledge and Kegan Paul for a book to be constituted by a lightly revised version of the thesis. As I continued to think about the subject I began to move in a radically different direction. The publisher patiently waited till a greatly revised version of the thesis emerged.

One prompt for this was what one might call the problem of the first person plural: what is it for you and me and possible others to become "we"? It was not clear that one could give an answer to this question in terms of the elements I had been considering. Indeed, individual expectations and preferences, albeit preferences and expectations regarding other people, and even common knowledge between individuals of such expectations and preferences, seemed to fall away into the background when this question was considered.

Charles Taylor had said something along these lines in the late 1970s and early 1980s. Durkheim had in effect said it in the late 1890s. In terms that incense some and inspire others, he says that a society or smaller social group is a synthesis *sui generis* of individuals. Thus you and I become "us".

To say any and all of these things is not of course to go very far. They present a problem: *what is it* for you and me to become "us"? What does the *sui generis* synthesis of individuals that Durkheim was talking about amount to? Could one reasonably talk about such a synthesis at all?

There are plenty of related problems, among them the following. What is it for people to act together—for an action to be *ours*, as they would say? What is it for *us* to intend to act in a certain way? What is it for *us* to believe something, as people often say that they do? What is it for *us* to have a social rule or convention? These questions can in principle be addressed separately, yet they seem to cry out for a unified treatment.

In a book published early in 1989, with the same title as my doctoral thesis but with from the same content, I offered a unitary theory of sociality at the level of "us". In subsequent writings I have continued to refine this theory and to argue for its relevance to different domains of inquiry, including moral and political philosophy. It will be useful to say something about it before proceeding further.

In summary, I have argued that for you and me and ... to become "us" all that is necessary is that we become parties to a particular *joint commitment*. Sociality at the level of "us", then, is achieved whenever a joint commitment is formed. By my stipulative definition, any population of jointly committed persons constitutes a *plural subject*. Hence I refer to the theory I have just described as *plural subject theory*.

As I understand it, Jack and Jill, say, have a joint commitment when they have a commitment such that, when someone asks "whose is it?" the correct answer is "theirs". And if one goes on to suggest "Are you saying that he is committed and she is too?" the answer is: "No, I am saying that *they* are committed: there is a commitment of the two of them that is not the conjunction of his personal commitment, on the one hand, and hers, on the other." A particular joint commitment is created when, roughly, each of the would-be parties openly expresses his or her readiness to be jointly committed in a specified way with the others, and this is common knowledge among them.

The number of parties to a given joint commitment can in principle be very large. I see this as having important ramifications for political philosophy, as explained in my book *A Theory of Political Obligation: Membership, Commitment, and the Bonds of Society* (2006), and elsewhere.

While I use "joint commitment" as a technical phrase, my proposal is that the concept it expresses plays a fundamental role in the life of human beings. This concept lies at the core of such central everyday concepts as those of *our* action, *our* belief, and *our* rule or convention. And the joint commitments people create bring into being the phenomena that instantiate these concepts. In this way these paradigmatically social phenomena are clearly *constructed by* human beings.

How precisely is the concept of a joint commitment incorporated in the everyday concepts mentioned and others like them? As an example, take the case of *our* belief or, as I shall also put it, *collective* belief. When we collectively believe that such-and-such, I propose that we are jointly committed to constitute as far as is possible, by virtue of the actions of each, a single body that believes that such-and-such. In other terms, we are jointly committed to *enact* such a body by virtue of the actions of each.

Why posit any kind of joint commitment here? One reason for doing so is that people take a set of obligations and rights to come into being when a collective belief comes into being. A joint commitment account predicts this because, as I have argued in several places, those who enter a joint commitment of any kind are in an important sense obligated to the other parties to conform to the commitment. One who defaults on such an obligation offends against the person he (or she) is obligated to as does anyone who fails to give someone what it is he has a right to.

Similar things can be said about our action, our intention, and so on. I have argued both *for* a joint commitment account of such situations and *against* various accounts that are *singularist* in nature. By a singularist account I mean one that appeals only to such features of single individuals as their desires, beliefs, commitments, and so on. An account like mine that appeals to a commitment that is irreducibly joint is, evidently, not a singularist one. Note that the concept of common knowledge is a singularist one, in that it appeals to what is known to or can be inferred by particular individuals, albeit about other individuals.

What topics should receive more attention in the future?

As said, I have argued in favor of a non-singularist account of collective belief and so on. Others, however, have developed singularist accounts of such things.

The question: what is it for *us* to intend to do something or, indeed, to do it, has probably received the most attention of all of those mooted here so far. Early in the 1990s, several theorists offered influential singularist accounts of these matters.

Thus Michael Bratman's very carefully delineated account of "shared intention" appeals at base to the personal intentions of the people involved. These intentions have to do with what will transpire as a result of the actions of all, but they are the intentions of single individuals. John Searle has proposed that "collective actions" such as an orchestra's playing of Beethoven's Fifth Symphony are constituted at least in part by "we-intentions" which are intentional states of particular individuals. According to Searle, a given person's we-intention is expressed when he sincerely says something of the form "We intend to do such-and such". We do something when, roughly, each of us acts on the basis of his or her own we-intention. A similar proposal was made by Wilfrid Sellars in the 1960s.

It appears from what they have written that Bratman, Searle, and most other contemporary theorists feel that unless one keeps within a singularist framework one will stray from the path of philosophical virtue. Such theorists often dangle before their readers the specter of a "group mind" or a "fusion of minds' which, they indicate, is not something any respectable theorist would have any truck with. Other common ways of casting aspersion on a view in this area are to characterize it as "supra-individualist" or "holist".

There may of course be philosophically suspect views that can reasonably be characterized as involving a "group mind" or a "fusion of minds", or as being "supra-individualist" or "holist". That does not mean that one must suspect any position that lies outside a singularist framework. Evidently I do not believe that the notion of joint commitment that I have invoked, for example, lies outside the pale of philosophical respectability.

There appears to be some kinship between my latter-day appeal to joint commitment and historical appeals in moral and political philosophy to a "common will", a "unified will" and the like. (Rousseau's appeal to "the general will" is different, having a special slant.) Precisely what was intended by the authors of these appeals is something on which I have little expertise. At least on the face of it they appear to express a sense that a non-singularist notion is needed if we are fully to describe the social world. It would be good to see further exploration of this historical material as a contribution to the philosophy of social phenomena.

So far philosophical social theorists such as Bratman, Searle, and myself have tended to focus on the development of our own accounts of, for instance, *our* intention. A given account may have been usefully clarified and highly refined over the years. That is fine as far as it goes, but the relative merits and demerits and, more generally, the relationship between the alternatives that different theorists have developed are seldom carefully addressed. When the views of others are discussed, it is often in a cursory manner, as part of the background to a theorist's own proposal. I should like to see more in the way of careful, detailed comparative assessments, not necessarily by the theorists in question themselves.

Sometimes it will turn out that different theorists are pursuing different questions. Some may be concerned with several questions at once. Thus Lewis, in his discussion of convention, had two different aspirations: to give an account of the everyday concept of a social convention, and (as fallback) to give an account of something that was an important phenomenon by any name. I believe, though, that there is often a genuine opposition between different views and hence a fruitful field for comparison and comparative evaluation. At least one common aspiration is to capture central everyday concepts such as that of *our* intention, acting together, and so on. Given that aspiration, I would like to see the question of singularism discussed further. A given singularist account may have described important aspects of the matter in question. But does it reach its core?

I have posed the question in terms of singularism versus non-singularism. It is likely that other technical terms and distinctions will need to be brought into play in comparing different theories with one another, and thus helping everyone to see the lie of the theoretical land as clearly as possible.

The debate about singularism versus non-singularism in relation to *our* intentions and actions may seem to lie in action theory. The same debate in relation to collective belief may seem to lie within epistemology. And so on. It is perfectly reasonable to claim that they do. At the same time these debates have a clear title to lie within the philosophy of the social sciences, if not within the social sciences themselves. Indeed, what we have here is not far from a replay, in contemporary philosophical guise, of a familiar debate: Weber or Durkheim? Which one has the most plausible type of account of the social realm?

What is the most important contribution philosophy has made to the social sciences?

Rather than attempting to pick out any particular contribution as the most important, I shall focus on one important kind of contribution philosophy can make to the social sciences.

I have said a fair amount about everyday concepts here, and about giving an account of a particular social phenomenon according to one or another everyday concept. I propose that such accounts – in particular, of course, the most adequate ones – are of great significance for the social sciences.

From the start, social scientists have differed over the importance of everyday concepts for their discipline. To take Durkheim first, he expressed a strong skepticism about everyday concepts and perhaps more importantly, everyday suppositions about the way things were. If sociology were to be a science, it must adopt empirical methods, at the least carefully observing what goes on in the world, gathering statistics, and so on. He had little respect for what might, today, be referred to as "folk sociology".

To some extent this must be right. Empirical generalizations made off-the-cuff are not necessarily to be trusted. They stand in need of confirmation – or disconfirmation – by more systematic inquiries. What of everyday *concepts*? As Durkheim says, these were framed for practical purposes, not for scientific ones. If one is to understand everyday lives, however, must one not attempt to articulate the concepts in terms of which they are lived?

Weber emphasized the importance of such understanding. In simple terms, human behavior divides into actions of different types by virtue of the intentions of the people in question, and intentions are couched in terms of particular concepts. In other terms, they have a particular content. Is this man chopping up a tree trunk in order to heat his house or in order to earn a wage, or impress a friend? That depends on how he sees the matter, how he would describe it. (Even an unconscious intention would need to be couched in terms of concepts possessed by the person in question.)

In spite of this emphasis, Weber was critical of some central everyday concepts. He insisted, for instance, that there was no such thing as a "collectivity which *acts*", whatever people seemed to think when they said things like "The orchestra played brilliantly". The everyday notion of a collective's action, then, could have a place only in describing the intentions of the individuals studied. It had no place in the set of concepts social scientists would use for their descriptions of the way things were.

Some everyday concepts may, of course, be suspect in various ways. They may even be logically incoherent. In order to know what the situation is with respect to any given concept, we need to understand what it amounts to. It may not be suspect at all. On the contrary, it may be well suited to inform the social scientist's own descriptions of the world (as well as his descriptions of the intentions and so on of the people who form part of that world). Pace Weber, the concept of a collectivity or social group which acts may be of this sort.

For several reasons, then, the social sciences would do well to consider accounts of central everyday concepts pertaining to the social domain such as those philosophers engaged in conceptual analysis aim to provide.

Of course, it is not an easy thing to give an account that is acceptable to everyone. This does not show that the philosophical enterprise cannot be of help to the social sciences. Insofar as philosophers are not yet in agreement, an understanding of the range of their accounts may be found useful. A given account may be chosen as the one to work with—with the caveat that it is not the only word on the subject and may turn out not to be the last one.

Many social scientists have in fact drawn upon the accounts of such philosophers as Bratman, Lewis, and myself in developing their theories. Michael Tomasello and his colleagues in developmental psychology are but one example. Examples can also be drawn from economics, political science, social anthropology, and yet other social sciences.

One might expect there to be some degree of cross-fertilization in such cases, with social scientists providing data or raising questions that philosophers can use as prompts for their own hypotheses. At the end of the day, a clear distinction between the philosophy of social phenomena and social science itself may be hard to make.

6
Daniel M. Hausman

Herbert A. Simon Professor
University of Wisconsin-Madison, USA

How did you get interested in the philosophical aspects of the social sciences?

In the political ferment of the late 1960s and early 1970s, I looked to both philosophy and to the social sciences to help figure out what was happening and what should be done. I grew increasingly interested in philosophical questions concerning the social sciences. In particular, I wanted to know whether they were "really sciences" and how much confidence one could place in them.

In the Fall of 1976, John Eatwell, now Lord Eatwell, gave a series of lectures at Barnard College concerning the so-called "Cambridge Controversy" in the theory of capital. This controversy focused on the possibility of "reswitching" and "capital reversing". Reswitching occurs when one technique of production T has a higher rate of return at both low and high interest rates, while another technique, T' has a higher rate of return at intermediate interest rates. Since a more capital-intensive production technique should be profitable when the rate of interest is lower, reswitching should not occur: T' obviously cannot be both more and less capital intensive than is T. Capital reversing occurs when less-capital intensive methods of production are employed when the rate of interest is lower. When reswitching occurs at least one of the changes of technique must be capital reversing, but capital reversing can occur without reswitching. Economists associated with Cambridge University, including Eatwell, argued that capital reversing showed that there was something incoherent in mainstream capital theory and indeed in mainstream microeconomics altogether. Economists located in Cambridge, Massachusetts argued in contrast that there was no evidence that reswitching or capital reversing ever actually occur and that in any case general equilibrium models have no need of any notion of capital or profits.

Eatwell was an exciting lecturer, and this controversy was full of drama and passion, while at the same time apparently touching on fundamental questions concerning economics. It was radically different from any of the models of scientific disputes that I had read about in my philosophy of science courses. So I thought to myself, "Why not write a dissertation in philosophy of science in which you attempt to understand what is going on in the Cambridge Controversy?"

At that point, I had a reasonably strong background in philosophy of science and philosophy of social science, thanks to the outstanding teaching of Isaac Levi, Sidney Morgenbesser, and Howard Stein at Columbia and to Hugh Mellor at Cambridge. But apart from a semester of introductory economics as an undergraduate and a smattering of Marxian economics, I knew nothing of economics. Several members of the economics faculty at Columbia helped me to cram in a reasonable grasp of microeconomics, general equilibrium theory, and capital theory, particularly Ronald Findlay and Martin Osborne.

In my dissertation, which became my first book, *Capital, Profits, and Prices: An Essay in the Philosophy of Economics,* I came to very different conclusions than Eatwell concerning the significance of the Cambridge Controversy, and my view of the overall structure of mainstream economics differed from the views defended by both sides in the controversy. In opposition to the critics who took the possibilities of reswitching and capital reversing as undermining mainstream economics, I agreed with the defenders that these possibilities revealed only the inadequacies of simplified aggregative models. But in opposition to the defenders of mainstream economics, I argued that economics cannot do without the notion of "capital" and its average rate of return. So the difficulties are, I believe, indeed very serious; and they are not resolved either by retreating to completely disaggregated general equilibrium models or (as some of the critics urged) embracing the neo-Ricardian approach sketched by Piero Sraffa in his *Production of Commodities by Means of Commodities.*

Though I focused on conceptual, as opposed to empirical issues, initially I was concerned with detailed questions concerning the precise content and assumptions of microeconomics, general equilibrium theory, and capital theory and with the character of debate and assessment within economics. Unable, as I was, to resuscitate capital theory, I focused thereafter on the methodological and philosophical questions to which my reflections of capital theory led, both specific and general. The most general problem concerned what to think of fundamental economic generalizations, such as the claim that people prefer more commodities to fewer or the claim that people's preferences are transitive. These generalizations are obviously not universal truths, and it is hard to see them as fundamental "laws", analogous to Newton's or Maxwell's Laws. Yet they appear to capture something real and to make possible scientific explanations.

When pressed to confront this problem, mainstream economists typically adopt one of two responses. The first denies that claims such as these are empirical generalizations at all. Instead they define *homo economicus*, perfect competition, or other models, and as definitions they are safe from empirical refutation. As it stands this response is unsatisfactory. Though there is a good deal of conceptual exploration in economics, and many models are not intended to be accurate representations of anything real, one cannot sensibly apply a model for the purposes of explaining, predicting, or guiding policies without taking its axioms as making true or false claims about the phenomena to which one is applying it.

The second main response is Milton Friedman's. He argues that it doesn't matter whether these generalizations are true or false. All that counts is whether the models economists employ that incorporate these generalizations make correct predictions. But this response is also unsatisfactory. The same facts that inform us that generalizations such as "people prefer more commodities to fewer" are not universal truths also refute predictions that are derivable from such generalizations. So Friedman's defense of what he calls "unrealistic assumptions" makes sense only if one can distinguish the predictions of economic theory that "count" from the ones that do not count. But as even Friedman recognizes (although not very clearly), the truth or falsity of implications of a theory that do not concern the phenomena in which economists are interested are not irrelevant. They help to guide modifications in models and to indicate where models are likely to work and where they are likely to break down.

Figuring out what one ought to say about such generalizations is harder than criticizing what economists have said. The approach that I pursued harkened back to a traditional view enunciated originally by John Stuart Mill in the first half of the Nineteenth

Century. According to this view, one should regard these generalizations as *inexact*. They are obviously not true universal generalizations, but they do capture genuine "tendencies" that are subject to the interference of "disturbing causes."

So my work on capital theory led to my general interest in economic methodology. That interest in turn led to my founding the journal *Economics and Philosophy* jointly with Michael McPherson. Editing *Economics and Philosophy* in turn got me involved in other areas in which philosophy and economics overlap, in particular questions concerning rationality, welfare, freedom, equality, and rights. My foray into capital theory also led to a continuing interest in causality, which was initially inspired largely by the work of social scientists, including especially Herbert Simon and Hubert Blalock.

Which social sciences do you consider particularly interesting or challenging from a philosophical point of view?

I find all of the social sciences fascinating, though I have only an amateur's knowledge of most. I have stuck to economics in order to make use of what I have already learned, not because I find it more interesting, more philosophically puzzling, or more valuable than the other social sciences. As the years have passed, the philosophical questions raised by the social sciences that have interested me most have changed. For a long while I was particularly interested in understanding causal explanations in economics and the other social sciences. Although some economists and social scientists (such as Milton Friedman) are inclined to deny that the social sciences have any interest in explanation, on the grounds that the ultimate goals are the practical ones of guiding policy, this is a mistake. Even if the ultimate goals are exclusively predictive, it is absolutely essential to be able to *diagnose* market failures, policy disasters, and anomalies of all sorts. Furthermore, for policy purposes, knowledge of correlations is not sufficient. Barometer readings are correlated with the onset of storms, but one cannot hold off a storm by jiggering with the dial on a barometer. One needs to know which correlations will break down when one attempts to use them to bring results about and which will hold up. So even if one's goals are entirely practical, one needs to explain phenomena and to identify causal relations.

In thinking through these issues, I came to realize not only how important it is to understand the nature of causation, but that the *asymmetry* of causation is absolutely critical. For example, the height of a flagpole (h), the angle of elevation of the sun (a), and the length of its shadow (s) are (in virtue of the law the light travels in a straight line) related to one another by a simple symmetrical function ($\tan a = h/s$). But one can use this relationship to explain only the length of the shadow, not the height of the flagpole or the angle of elevation of the sun, and if one wants to change the angle of elevation of the sun or the height of the flagpole, this relationship is of no help. Causes explain their effects, but effects do not explain their causes and effects of a common cause do not explain one another.

So the direction or the asymmetry of the causal relation is absolutely crucial, and for nearly twenty years a great deal of my philosophical research focused on attempting to understand that asymmetry. Merely to insist, as Hume did, that causes precede their effects in time, fails to distinguish between genuine causal relations and relations between effects of a common cause where one precedes the other, and the temporal priority of cause to effect does nothing to make sense of the fact that causes explain their effects while effects do not explain their causes. It turns out that there are a number of different features of causal asymmetry, and my book *Causal Asymmetries* explores the relations among these different asymmetries of causation as well as arguing that a particular asymmetry of "independence" is fundamental.

Over the last few years my interests have shifted, and I am currently especially interested in understanding how people evaluate alternatives. Evaluation seems straightforward in the case where alternatives have purely instrumental value and the value of their possible consequences is already known. But apart from that special case, it seems that little is known and little has been said about evaluation. In Hume's view, there is nothing normative to be said. "Reason is, and ought only to be the slave of the passions." So the passions, including evaluations of states of affairs, cannot be reasonable or unreasonable. Indeed, Hume maintains "Where a passion is neither founded on false suppositions, nor chuses means insufficient for the end, the understanding can neither justify nor condemn it. 'Tis not contrary to reason to prefer the destruction of the whole world to the scratching of my finger."

But Hume's view is hard to accept. Consider the contemporary work concerning the evaluation of health states (which are the source of my interest in the subject). How, for example, should one go about deciding whether it would be worse to be confined to a wheel chair or worse to be deaf? This seems to be a cognitively demanding question that an individual might answer correctly or incorrectly, not just an occasion to consult one's gut. The Humean view that desire or preference is not subject to reason seems unacceptable. But how should one go about rationally deciding on an answer?

At this point I still do not have any good answer. My work has mostly been critical of the Humean view and of contemporary efforts to evaluate health states by measuring preferences, which are still very much in the spirit of Hume's view. The path toward developing an alternative leads through a general substantive theory of prudential value and an account of how public deliberation can secure a commitment to particular modes of evaluation.

How do you conceive of the relation between the social sciences and the natural sciences?

I am assuming that what is being asked concerns my views of the similarities and differences between the social and natural sciences. The vague general question of "social scientific naturalism" – whether the social sciences resemble the natural sciences in their goals, methods, and concepts – is what initially got me interested in the social sciences. But it is not a very well-formed question. At the level of day-to-day practice, the many specific social sciences differ dramatically from the many specific natural sciences. Social scientists use different instruments, study different objects, employ different concepts, and have different specific objectives. But of course the specific social sciences also differ dramatically from one another – as do the specific natural sciences. On the other hand, at the most general level, the social sciences and the natural sciences are much alike: Both attempt to gain knowledge of their subject matter, and in doing so they are sensitive to empirical evidence.

So there is a real problem even making clear what is being asked when the question of the relations between the natural and social sciences is posed. I have a lengthy essay on this subject coming out soon in *The Handbook of the Philosophy of Economics* that Uskali Mäki is editing for North-Holland, and the issues are hard to summarize briefly. If one accepts some general philosophical characterization of any aspect of the natural sciences, one can ask whether that characterization is apt with respect to the social sciences. In addition, one can look to the history of discussions of social scientific naturalism to see what issues previous authors have found central. I classify the questions under the following five headings:

1. *(Goals) If one is a realist or an instrumentalist concerning the natural sciences, does it follow that one must be a realist or an instrumentalist about the social sciences or vice versa? Do social scientific inquiries have additional or different objectives?* Although many prominent 20th-century commentators have defended the view that the ultimate goals of the social sciences should be exclusively practical, it is hard to see any reason why the quest for explanation would be central to the natural sciences, while misguided or futile in the social sciences. On the other hand, as Max Weber pointed out, social and psychological phenomena matter to people in ways that natural phenomena do not. Though some people may be interested in the history of a particular cave or turtle, the peculiarities of non-human individuals do not engage our passions and curiosity in the same way that the biographies of individuals, groups, and nations do. One important goal of social inquiries is consequently to provide such narrative information concerning particulars.

2. *(Objectivity and values) Can the social sciences be objective or "value free" in the same way that the natural sciences are objective or value free?* In practice, values (beyond those which are implicit in all scientific practices, such as not forging data or sabotaging the work of competitors) are bound to influence work in the social sciences much more than work in the natural sciences. The reason is easy to see: the findings of the social sciences frequently bear directly on people's interests. Marx does not exaggerate too much, when he refers to "the Furies of Private Interest." Although these practical difficulties are often very serious, they are arguably not of much philosophical interest and do not mark a difference in kind (as opposed to a difference in degree) between the natural and social sciences. The direction, funding and findings of research in the natural sciences affects people's interests, too.

There are also some ways values influence work in the social sciences that do not arise in the natural sciences. One of these has already been touched on. People are interested in and engaged with the peculiarities of people's lives, and these interests clearly depend on evaluative commitments and interests. Furthermore, values influence not only social investigators, but the objects they investigate. Though there seems to be no principled reason why value-free inquiry concern values is impossible, it is bound to be difficult. Values that differ from those of the investigators will be harder to understand, and their influence and robustness will be harder to gauge.

3. *(Testing) Can theories in the social sciences be tested and confirmed or disconfirmed in essentially the same way that theories in the natural sciences are appraised? For example, is introspection a special source of knowledge of psychology?* Many theories in the social sciences are tested and confirmed or disconfirmed in just the same way that theories in the natural sciences are. So clearly they can be. There are nevertheless at least three differences. First, social scientists have access to large data sets that are typically designed by government agencies rather than by social scientists themselves and thus measure different things than social scientists would like to measure. Problems of inferring what one wants to know from masses of tangential data are much more common in the social sciences than in the natural science. Second, experimentation in the social sciences faces much more confining ethical, financial and technical limits than experimentation in most of the natural sciences. Third, introspection or envisioning some social circumstances or process "from the inside" apparently provides an additional source of data, which is unavailable in the natural sciences. Although not trivial, none of these differences seems however to mark a fundamental divide between the natural and social sciences. The differences between testing theories in astronomy and testing theories in neurology seem to be as large as the differences between testing theories in organic chemistry and testing theories of social structure.

4. *(Reduction and ontology) Are social entities such as norms, cultures, institutions, tribes, or classes "real"? Are they reducible to physical things?* These are fascinating questions for philosophers of the social sciences and for social scientists who are inclined to philosophical reflection, but I do not think that they mark a fundamental divide between the natural and social sciences. Even if one concludes that social entities are in some sense not "real," and one denies that they are reducible to physical things, they are obviously not just fictions, on a par with gremlins or hobbits. Corporations buy and sell things. Baseball teams win and lose games. Languages evolve from one another. Though these claims may need lots of reinterpretation, they can hardly be denied. And that seems to leave room for generalizations, testing, predicting, and so forth.

5. *(Explanation)* Do the same models of explanation apply to both the natural and social sciences? Many explanations in the social sciences involve reasons and norms. How are explanations that cite reasons or norms related to causal and theoretical explanations in the natural sciences? It seems to me that the fact that the most common forms of explanation in the social sciences cite people's reasons for their actions distinguishes them significantly from explanations in the natural sciences. Not all explanations in the social sciences cite reasons. For example, Brian Skyrms in his *Evolution of the Social Contract* employs evolutionary game theory to show why the strategy of sharing unowned goods equally (as in "divide the cake problems") will be selected for. Whether or not his story is a good one, it does not rely on any claims about the reasons why agents tend to "divide the cake" equally. But most explanations in the social sciences cite reasons, because a particular "folk psychological" theory of choice is deeply entrenched. According to this theory, one explains actions by citing an agent's beliefs and desires. These in turn either state or point to the agent's *reasons*, and reasons are subject to evaluation. They may be justified or unjustified, and their justification or lack thereof is typically conferred on the actions to which they lead.

In the 1950s, under the influence of Wittgenstein, a number of philosophers argued that explanations in the social sciences, citing as they do reasons and norms, are not causal explanations at all, and that explanation in the social sciences is completely different from explanation in the natural sciences. These views are exaggerated and untenable. Whatever appeal they may have had at the time was tied to the general qualms philosophers and social scientists felt about causal notions and the consequent effort to reduce talk of causation to talk of laws. Though general laws are in short supply in the social sciences, there are no difficulties applying more sensible views of causal explanation. Whatever one wants to say about reasons should not lead one to deny that the blaring of the midnight fire alarm caused the hotel residents to leave the warmth and comfort of their beds. And as Donald Davidson famously argued in the early 1960s, it seems that those who cite reasons to explain actions have to invoke causal notions to distinguish which of the many reasons an agent may have had for a particular action was responsible for the action.

Although explanations that cite reasons are thus a species of causal explanation, they are nevertheless a special and distinct species. The fact that reasons are subject to evaluation matters.

Bad reasons are harder to accept. Claims about which reasons motivate people are claims about the reach of morality and the extent of altruism. They express and influence attitudes toward other people. They have the potential to provide (or to undermine) a moral education. Moreover, when social investigators cite an agent's reasons, they have to describe the phenomena using the terms and concepts that the agent uses. Regardless of what they may think about unconscious or even neurological factors that influenced the action, if they want to understand the reason, they need to be able to describe what is at issue in the terms in which the agent described it. And in some cases, finding that description – which task rarely arises in the natural sciences – may be the hardest part of the explanation.

At the end of the day, explanations of particular happenings provided by the social sciences are typically causal explanations. But lots of the day passes before the end, and much of it is filled with the tasks of describing social circumstances in the terms in which they are understood by the participants so as to be able to formulate the agent's reasons and then evaluating those reasons.

When the discussion is finished – and obviously I've only scratched the surface – should one conclude that social scientific naturalism is true, that the social sciences are fundamentally "like" the natural sciences, or should one conclude that it is false? I am not sure, and I doubt that it matters. The substance resides in answers to the more specific questions sketched above.

What is the most important contribution that philosophy has made to the social sciences?

The social sciences only emerged out of philosophy fairly recently, and the major advances brought about by figures such as Adam Smith, John Stuart Mill, Karl Marx are scarcely separable from their philosophies. Even in the late Nineteenth and early Twentieth Century, the great social theorists were heavily involved with and interested in philosophy. It is very hard to pick out the most important contribution of philosophy. Is it the idea of a probability distribution? Or a theory of meaning? Is it the notion of the experimental method? Is it the notion of intentionality, of a functional explanation, of unintended consequences, of a mechanism, or of causal graphs and structural equations? And at the same time that philosophy has contributed to the social sciences, it has also sometimes harmed them. The radical empiricism of the logical positivists, which was tied to their qualms about invoking causal relations and their denial that value judgments have cognitive significance probably set the social sciences back significantly.

Recent philosophical work on causation and causal inference has, I believe, the potential to make a very significant positive contribution to the social sciences. As pointed out especially by Peter Spirtes, Clark Glymour and Richard Scheines in *Causation, Prediction, and Search* and related work, causal structure has testable implications. With the help of some ineliminable and fallible causal assumptions, one can draw causal conclusions from data even when experimentation is impossible. Indeed Spirtes, Glymour and Scheines have even developed computer software to automate the search for causal relations. This work is very exciting both for the technology it provides and for the rehabilitation of specifically causal inquiry that it promises.

Which topics in the philosophy of social science will, and which should, receive more attention than in the past?

As already mentioned, I think that more attention should be paid to the question of how individuals separately or collectively evaluate alternatives. On a practical level, it seems to me that by far the most pressing problems are collective action problems. If one looks around, the really enormous problems people face are all social problems. This is true even in a case such as the AIDS epidemic in which the culprit is a pathogen and the ultimate cure medical. If we better understood social relations, we could dramatically limit the spread of the disease, and we would bring the drugs that currently control the disease to millions of people who instead are going to die soon of the disease. Though there are technological aspects of the other scourges of our times – ethnic and religious conflicts and terrorism, global distribution, poverty and famine, and environmental degradation and global warming – they are all fundamentally social problems. And the greatest hurdles that need to be jumped to solve them are problems of cooperation and social organization. If human beings manage to destroy their environment, extirpate themselves, or simply bring about an era of terror, chaos and misery, it will be for want of knowledge or lack of ability to forge institutions that act in a way that is collectively rational.

7

Harold Kincaid

Director, Center for Ethics and Values in the Sciences
Professor of Philosophy
University of Alabama at Birmingham, USA

How did you get interested in the philosophical aspects of the social sciences?

I entered college in 1970 when the civil rights movement and the antiwar movement were in full swing in the US and I was involved in those movements in various ways. I also grew up with considerable direct and eye opening experience with the US mental health system and mental illness because my father was a manic-depressive who took his life right before I entered college. Then after college I spent four years as a union reform activist while working for AT&T at a grindingly awful job. I got a first hand look at the downsides of bureaucracies and at systemic discrimination in the work place (against women). So I entered graduate school with a strong interest in social problems and in what the social sciences could say about them. My undergraduate degree was in philosophy, though I had extensive course work in the social sciences. My main mentors had recently finished dissertations with Quine and Goodman and were early defenders of nonreductive physicalism (Hellman and Thompson 1976) and had an interest in the philosophy of the social sciences, especially economics. My graduate degree required a minor and mine was economics. So it was natural for me to take the naturalism and philosophy of science I had gotten from studying Quine and Goodman and apply it to the social sciences; my naturalism led me to believe that philosophy of the social sciences is continuous with social science itself.

Which social sciences do you consider particularly interesting or challenging from a philosophical point of view?

The question rests on a presupposition that I think is highly dubious—that there are multiple, independent social sciences. That may be still true as institutional fact, but I do not think that it reflects any real division in the social world as it were. There is no strictly separating the economic, social and political, though there may be times where there is some reasonable sorting of causes into economic vs. political, for example. I take this to be an open interesting question in the philosophy of the social sciences and believe that recent developments in economics bringing institutions back into the analysis argue against any very useful separation.

One aspect of the social sciences that I have long found philosophically challenging is the nature of social explanation and the relation of large scale explanations to explanations of individual behavior. In effect, my interest has been in extending certain versions of nonreductive physicalism first developed in the philosophy of mind to understanding the relation between the social and the individual in social science explanation. My two books in the 1990s pursued that project (Kincaid 1996;1997).

The aspect of social science that I currently find particularly challenging concerns how social scientist can provide evidence for the kinds of complex causality that social sciences must face up to if they are going to progress. Much in the social sciences has long centered around the equations and variables approach tested via multiple regression. That approach rests on some philosophy of science assumptions I find questionable. It presumes a simplified notion of causality, e.g. that social causes have separate, independent and constant effects. Second, it expresses a faith in a formal logic of inference.

Both presuppositions seem to me dubious and an obstacle to successful social research. Potential causes in the social realm need not be operative everywhere, they may be necessary causes only, they may not have separate effects independent of the presence and level of the other factors. These kind of causes are not easily if at all representable in standard linear regression equations. Real social processes are frequently path dependent, the result of tipping points and binding constraints. It was easier to ignore this in areas like economics so long as economic variables were thought to be a nicely separable realm, but as economics has developed to bring in the roles of institutions, norms, evolutionary processes, etc. that assumption is often not valid.

The formal logic of inference ideal manifests itself in the wide spread practice of assuming that the logic of probabilities suffice for making inferences in the social sciences. That assumption is controversial for at least two reasons. First, it is not clear what grounds the probability claims. Standard social science drops and adds variable based on tests of statistical significance. But those tests require a probability model—some analogue of sampling from a population. However, much social science data is not a random sample in an obvious sense nor the result of conscious randomizing, so it is not clear what statistical significance tests mean. There is some nod towards the idea of hypothetical populations in econometrics, but no detailed defense and elaboration is to be found. The second problem is that even if the probability foundations were clear, social scientists seem to think they can do much more than they really can. Statistically significant results are treated as probably true and insignificant results as probably false. However, statistical significance only tells you the probability of seeing the evidence by chance alone. However, knowing that is only part of getting the evidence for what we want to know, viz. given our evidence, how plausible is the maintained hypothesis?

The upshot here is that a dominant picture of causes and evidence in the social sciences is at odds with the real nature of causation and the kind of available evidence that might be used to assess it. So to me the interesting question is then: How do or how can the social sciences go about dealing with these realities? How well do philosophical accounts of causation and of evidence illuminate the situation? These issues are particularly important in areas such as development economics and political sociology where the goal is to infer the causes of differences in outcomes in systems with complex causality based on observational evidence for which there is no obvious probability model.

How do you conceive of the relation between the social sciences and the natural sciences?

I see no fundamental distinction between the natural and social sciences. Society is part of the natural world and is amenable to study with the methods of scientific inquiry for studying the natural world. Those who want to advocate the idea that the social realm must be studied by special methods suffer from a failure of nerve—a failure to face up to the fact that humans are not somehow special and outside the natural order. Of course, "amenable," "methods of scientific inquiry," and so on are vague; an interesting naturalism that denies a sharp distinction between the natural and social sciences has to give them some real content. One might describe the methods of scientific investigation so vaguely that no one could disagree with naturalism, e.g. as "reasoned investigation." On the opposite end of the spectrum naturalism could be made trivially false by being overly concrete. It is certainly true that social organization is not usefully studied by the use of microscopes. To get a useful thesis with content we need to recognize that scientific methods can be described at different levels of abstraction—"reasoned investigation" is too abstract and "uses a microscope" is too concrete. Moreover, there are scientific standards for different aspect of scientific practice, e.g. explanation or confirmation.

I would argue that we can get some traction if we focus on rigorous testing of causal explanations. I certainly do not think that either can be captured in purely formal terms or applied algorithmically. However, in practice we have lots of knowledge about what is involved. So there is a lot we could say about rigorous testing. Rigorous tests involve among other things (1) fair tests that do not one way or the other bias the case against competing hypotheses, (2) independent tests where the theory of the test is evidentially independent of the hypothesis at issue, and (3) cross testing where the hypothesis at issue is tested in evidentially different situations. Causation can often be shown when we use background causal knowledge in connection with evidence about correlations.

These are standard practices of both the natural and social sciences I would argue. They span the usual experimental vs. observational and quantitative vs. qualitative divide. Neither experimental evidence nor quantitative evidence is compelling if it does not produce a rigorous test and observational and qualitative evidence can be compelling under the right circumstances. No doubt the social sciences have a harder time getting the circumstances right than some natural sciences but then some good natural science, e.g. parts of ecology and evolutionary biology, face very similar complexities in rigorously testing causal claims as the social sciences often do.

A very traditional arguments for a sharp distinction between the natural and social sciences turns on the meaningful nature of human behavior. The claim is that understanding meaningful behavior involves entirely different practices than those of the natural sciences. Taylor's (1971) influential article puts it clearly: there are in the "sciences of man" no brute data like we find in the natural sciences. However, that picture of the natural sciences is no longer tenable (nor was it when Taylor advocated it), for we know that the relevant data, how those data are described and interpreted is not something we just read off from nature or do in so in a way dictated by the logic of science without reliance on background knowledge. Once we realize this, then any automatic sharp distinction between the natural and social sciences for these reasons looks much less plausible. Not surprisingly, if we look at natural sciences such as primatology or any areas that study animal signaling, we find that they have to deal with problems of meaning and that they do so by trying to conduct fair, independent and cross tests. To persist in claiming that interpreting human behavior is still something fundamentally different is, as I said before on my view, a failure of nerve—an inability to face up to the fact that humans are part of nature and not special creatures by virtue of having souls or some other metaphysically special essence.

What is the most important contribution that philosophy has made to the social sciences?

To be honest, I think the influence of philosophy on the social sciences has been of mixed value. Let me start with the more questionable influences. I noted above that much social research has been conducted with specific and implausible assumptions about the nature of evidence and causation. Also along with those assumptions went the idea that explanation is about identifying laws and often a general reluctance to talk about causes at all. It is not implausible to think certain understandings of positivism played a major role in the social sciences taking these stances. The regression equations and statistical tests approach was a real contribution in that it put constraints on the usual vague theoretical claims and informal evidence characteristic of the founding fathers of sociology, for example. However, as I argued above, it can also be a serious obstacle.

Philosophy from the continental and quasiWittgensteinian traditions has had some serious influence but again of mixed value on my view. The emphasis on social meanings, social context, norms, and the like has been positive in that it encouraged social scientists to look at these things, though arguably much of this is already there in Durkheim and Weber. However, this influence has tended to come with explanatory and epistemological baggage that I find implausible and detrimental to successful social science. The epistemological assumptions are illustrated by Taylor's defense of social science as beyond the pales of natural science standards that I mentioned above. The questionable assumption about explanation is that natural science explains via laws but a social science must be different. However, I would point to molecular biology as a paradigm of good science where laws are scarce and the focus is on causal processes, where causality is complex and contextual. So I think this aspect of the interpretivist tradition was also a misstep.

On the more positive side, David Lewis' (2002) work on conventions obviously had an important part in the development of game theory, which has been one of the more productive technologies in the social sciences. Philosophical work on decision theory and rationality played a part in the development of rational choice theory as well and to the extent that rational choice theory has been fruitful its impact has been positive. Feminist philosophers of science such as Alison Wylie (2002) have been helpful in bringing feminist insights to methodological discussions in some social sciences. Recent work on causality by philosophers such as Nancy Cartwright (2007) is finding its way into economics as is work on the methodology of experimental economics by philosophers (Guala 2005). In general the philosophy of economics has been moving toward the kind of concrete, scientific informed analysis that philosophy of biology and physics have been providing for some time and that has had a clear impact on those disciplines.

Which topics in the philosophy of social science will, and which should, receive more attention than in the past?

On my view, philosophy of the social sciences should follow the path of philosophy of biology and physics and focus its efforts on specific problems raised by the sciences with the aim of contributing something to our understanding of science and contribute something to clarifying the issues at stake in the sciences themselves. The traditional philosophical project of analyzing concepts in terms of necessary and sufficient conditions tested against linguistic intuition and producing more or less a priori conceptual truths is largely bankrupt as a tool for understanding science, though some useful skills and sometimes some clarity can result from such an enterprise. However, to offer a social science explanation, tradition dies hard and philosophers may have vested interests in continuing these practices. So I will not be surprised to see this tradition continue and the metaphysics of the social world and social meaning revealed from the armchair to retain its place.

So one answer to the question of where philosophy of social science ought to go is it should not go there—it should not pronounce on the scientific status of entire bodies of research but instead should use philosophical tools to engage in ongoing debates of relevance to social scientists. So there is much to be done, for the social sciences are rife with interesting controversies and problems to which philosophers can contribute and likewise learn useful things of philosophical relevance. Here are some suggestions that currently interest me (I won't pretend to predict who in the future besides myself will pursue them):

1. As I pointed out above the social sciences confront complex causality with observational evidence. Workers in macrosociology, political sociology, international relations, and development economics find themselves talking about necessary causes that may have no independent effect on their own, about tipping points and binding constraints, about processes acting on different times scales, and so on. They also find themselves using case study comparisons and small N studies. These kinds of causes and data are not easily fit into frameworks based on finding structure hidden in correlations or notions of causality that require causes have separable effects. Social scientists have done some interesting and creative work using Boolean algebras, fuzzy sets, and other methods to say something about how this all might work. (Abbott 2001; Ragin 2000). Getting clear on the strengths and pitfalls of this kind of research can help contribute to better social science.

2. Experimentation has found a growing audience in social sciences like economics. There are host of issues there where philosophers of science might contribute to clarifying practice and to increasing philosophy of science's understanding of experimentation. One key issue that has not gotten the philosophical attention it deserves is the problem of external validity – whether experimental results tell us about behavior outside the lab – and its relation to evaluating experimental results themselves. Results from experiments in a lab with undergraduate subjects may not tell us much about economic behavior in real markets. Clarifying what grounds successful inferences to outside the lab would be very useful. A re-

lated question is to what extent experiments themselves can go on without a good understanding of how the factors of interest work outside the lab. It is usually assumed that internal and external validity are independent issues. Yet knowing that there are factors in the real world not captured in the experiment might also make you wonder if those factors could not be confounding variables in the experimental situation that ought to be controlled. If social context, for examples, influences the game that people take themselves to be playing outside the lab, then there is the worry that games played in experiments are not the ones experimenters believe themselves to be investigating because social context is not controlled for.

3. The social sciences have had a long and ambivalent relationship to Darwinian type competition and selection explanations applied to social entities, practices, etc. Recent developments in evolutionary game theory, cultural selection models, mimetics, and so on have put these explanations back at center stage. There is much for philosophers of science to do here, and two decades of research in philosophy of biology can be of help. Much of this work gets its plausibility from optimality arguments, e.g. practice P would be an optimal way to achieve X, therefore it exists because of doing X. When are these arguments plausible and when are they Panglossian? How do issues about levels of selection arise when the social sciences provide them selectionist account? How can selection work without any clear parallel to the notion of genes? What is the proper place of rational choice explanations versus learning and cultural selection or, in other words, of the place of rational choice game theory and evolutionary game theory?

4. Integrating social explanations and explanations in terms of individuals. Individualism as a reductionist thesis I think is of little interest. However, weaker forms calling for mutual compatibility or, even better, explanatory integration are of interest. The questions on my view are empirical ones where conceptual clarification can nonetheless help. In particular, combining recent advances in cognitive science – cognitive psychology and cognitive neuroscience – with understandings of social process provided by the social sciences seems to me very important. Cognitive scientists often treat social causes as a separate factors or forces from the psychological and biological and definitely tend to think "their" causes are the most important. I think this confuses differences in explanatory tools with differences in the ontology of causes. The social factors affecting addiction, for example, are not

acting independently of biology nor are the biological ones acting independently of individuals in a social context. Finding coherent theses here is essential but there are many interesting questions in the background. One important idea of contemporary cognitive science is that minds and meanings are embedded in the environments around them—what does this claim come to and what does it say about the relation between the social and psychological and biological sides of our explanations of human behavior? I am not calling for more twin earth thought experiments but for investigation of specific biological and psychological accounts and how they relate – or should be related – to social explanations. Different versions of innateness hypotheses and the evidence for them is a related question that philosophers are beginning to pursue. Understandings of mental illness on these various dimensions is another vein that could be fruitfully mined.

5. The place of values in social science remains important. However, it not one issue but many (see Kincaid et al 2007): what part of social science is involved? what kinds of values are involved? are they essential? do they prevent objectivity? What are the social processes typical of the social sciences themselves? Detailed case studies analyzing these issues can teach us a lot about the social sciences, can help keep the social science honest, and can help develop a more sophisticated understanding of facts and values in science in general than is found in claims that values are never legitimately involved or that science is values and politics all the way down.

References

Abbott, Andrew. 2001. *Time Matters*. Chicago: University of Chicago Press.

Cartwright, N. 2007. *Hunting Causes and Using Them*. Cambridge: Cambridge University Press.

Guala, F. 2005. *The Methodology of Experimental Economics*. Cambridge: Cambridge University Press.

Hellman, Geoffrey and Thomson, F.W. 1976. "Physicalism: Ontology, Determination, and Reduction," *Journal of Philosophy* 72: 551–564.

Kincaid, Harold .1996. *Philosophical Foundations of the Social Sciences*. Cambridge: Cambridge University Press.

Kincaid, Harold. 1997. *Individualism and the Unity of Science.* Lanham, MD: Rowman and Littlefield

Kincaid, H., Dupre, J. and Wylie, A. 2007. *Value-Free Science: Ideals and Illusions.* Oxford: Oxford University Press.

Lewis, D. 2002. *Conventions: A Philosophical Study.* London: Blackwell.

Ragin, C. 2000. *Fuzzy Set Social Science.* Chicago: University of Chicago Press.

Taylor, Charles. 1971. "Interpretation and the Sciences of Man," *The Monist* 25:3–51

Wylie, A. 2002. *Thinking Through Things.* Berkeley: University of California Press.

8

Daniel Little

Chancellor and Professor of Philosophy
University of Michigan – Dearborn, USA

What brought you to the philosophy of social science?

My interest in philosophical issues in the social sciences came early in my graduate education at the Harvard philosophy department in the early 1970s. I had studied the philosophy of science and the philosophy of psychology as an undergraduate at the University of Illinois, and I was particularly excited by the work of Noam Chomsky on the intersection between language and cognition. As I left college I also began to develop a curiosity about Marx's theories, and in the first year or so as a graduate student I began reading Marx's work. As a second-year graduate student I organized a pair of undergraduate tutorials on the early Marx and the later Marx, which was a very good way to learn Marx's work. And I served as a graduate assistant in John Rawls's courses on political philosophy, in which Rawls gave several weeks of attention to Marx's early writings. (Rawls appeared to have minimal interest in the "social scientific" Marx of the later writings.) During graduate school I did a great deal of reading under my own direction of the classics of social theory—Weber, Durkheim, Simmel; also Marxist historical writings (Albert Soboul, E. P. Thompson, Eric Hobsbawm, Maurice Dobb, M. I. Finley); and such historical theorists as Immanuel Wallerstein. These sources gave me a deep fascination for the complexity of history and social processes and the variety of theoretical approaches that were possible for understanding this complexity. I had read much of the existing literature on the philosophy of social science (chiefly represented by the May Brodbeck anthology), which was largely inspired by a fairly positivistic and a priori approach to the subject. This approach I had found uninteresting, and incomparably less interesting than the specific empirical and theoretical work I was finding in the social sciences and in the classics of social theory. By the midpoint of my graduate studies I had arrived at these foundational questions, which continue to captivate me: What makes a study of social phenomena "scientific"? What theoretical tools are available to serve as a basis for explanation in the domain of social outcomes and processes? And how can philosophers play a contributing role in arriving at intellectually powerful and empirically supportable insights into the causes of social and historical change?

When it came time to write a dissertation I decided to focus on Marx's economic writings, and I defined my subject as a study of the assumptions about social scientific knowledge that Marx made in *Capital* and *Theories of Surplus Value*. I chose to focus my efforts by probing the epistemological frameworks and assumptions about social science that Marx made in his researches from the 1850s through the publication of *Capital* (1867). And I chose to approach the genre of the philosophy of social science, not through a canon of philosophers' writings, but through a concrete study of the scientific practice of one important founder of the social sciences, Marx. This meant posing new questions rather than following a script for analyzing a given set of philosophical problems. The result was a dissertation called *Marx's Capital: A Study in the Philosophy of Social Science*. The dissertation defined the problem as one of surveying the kinds of knowledge and explanations Marx attempted to offer of the social world and of the capitalist mode of production. The dissertation set its objective in these terms (the first words of the dissertation): "This thesis is an essay in the philosophy of social science. It is an attempt to address Marx's social theory as an important episode in the history of social science, and to try to uncover in detail its implicit standards of rational scientific practice.... It is important to try to discover the epistemological and methodological characteristics which define it, or in other words, to discover in detail the standards of empirical rationality which underlie its scientific practice." The dissertation pointed out the importance of constructing philosophy of social science out of direct engagement with rigorous pieces of social scientific reasoning, and it rejected the idea of constructing philosophy of social science out of analogy with the natural sciences.

The dissertation was written just before the flourishing of a new current within Marxist theory, analytical Marxism. The dissertation was accepted in 1977, and it was my intention to publish it as a book. However, the first valuable writings of the analytical Marxism were beginning to appear at that time—John McMurtry and Gerald Cohen in particular—and so I chose instead to re-think my findings and to formulate a new set of questions in light of the new discussions emerging. The result was *The Scientific Marx*, which was aimed at identifying the logical and methodological issues that genuinely mattered in assessing Marxism as a basis for social science research and theory. Several points were particularly significant: that Marx's treatment of capitalism does not amount to a unified theory, that there is no "dialectical method" at work, and that Marx's explanations are generally understandable along the lines of a "logic of institutions," in which the researcher identifies a set of institutional opportunities and constraints and works out the aggregate consequences for social outcomes when large numbers of prudent agents work within these institutions.

After completing *The Scientific Marx* I became interested in a series of debates that were occurring in Asian studies, particularly the "rational peasant" debate, the "moral economy" debate, and the "Chinese stagnation" debate. I undertook to analyze these extremely interesting discussions within the literature of Asian studies, using some of the analytical insights of the philosophy of social science. I found that scholars of Asian studies were particularly receptive to the incursions of a philosopher, and I took great benefit of the intellectual generosity of such scholars as historian Paul Cohen, political scientist James Scott, and anthropologist G. William Skinner. Jon Elster's book, *Explaining Technical Change*, served as an excellent model for me of a philosopher's approach to important issues in the logic of social science and history through careful study of several interesting cases. This research on debates in the China field resulted in a book called *Understanding Peasant China: Case Studies in the Philosophy of Social Science*. Since then my work has focused on issues having to do with the logic and force of social explanation, and with the ontology of the social world. I have always tried to engage the philosophical issues by taking seriously the research and knowledge created in various areas of the social sciences. Interaction with working political scientists, anthropologists, sociologists, geographers, and Asian studies specialists has been an enormously fruitful part of my intellectual development.

A consistent and defining intuition that has guided my explorations of the logic of the social sciences is a deep dissatisfaction with the positivist philosophy of social science of the 1960s and an abiding fascination with the work of innovative, rigorous historians and social scientists. The idea that positivism defines scientific rigor and the structure of scientific knowledge is a deeply unconvincing one when applied to the social sciences. The chestnuts of positivist philosophy of science—the covering law model, the search for unified theories of a domain of phenomena, the distinction between observation and theory, the hypothetico-deductive model, and the doctrine of the unity of science—do a bad job of measuring or illuminating the sociological imagination and the rigor of social science reasoning. Scientific method needs to be suited to the nature of the phenomena for which the science is constructed. The social world is variegated, heterogeneous, contingent, multi-causal, and plastic, and therefore the methods and theories of the social sciences need to be constructed with this heterogeneity in mind. The skills of a biographer, a medical diagnostician, a forensic engineer, or a literary critic are more relevant to the social sciences than those of an empirical chemist or a theoretical physicist. Social scientists are indeed rigorous and ingenious in their methods of probing the workings of social systems and structures; but we need to discover the specific features of the rigor that they achieve by examining their work in detail.

A concrete example of the collaborative engagement between philosophy and social science that I favor took the form of a two-year research fellowship from the Social Science Research Council under an innovative program called "International Peace and Security." The idea of the program was to encourage scholars outside of international relations to take a substantive interest in some aspect of international security. My topic was "Food Security and International Development," and I spent two years (1989–91) at Harvard's Center for International Affairs learning quite a bit of development economics and development theory. Particularly valuable were seminars and conversations with development economists at the Harvard Institute for International Development, area specialists on Asian development at Harvard and MIT, and practitioners of development policy in a variety of research organizations. Among the talented social scientists whose work stimulated my development as a philosopher of social science during these two years were Atul Kohli, Dwight Perkins, and Peter Timmer. This experience led to a different kind of interaction between philosophy and the social sciences—this time focused on issues of theoretical adequacy and normative analysis. How good a fit is there between development theories and the actual economic and cultural experience of developing countries? And what role might theories of distributive justice play in the design of development policies? The first set of interests led me to conceptualize a volume in the philosophy of economics called *On the Reliability of Economic Models*, in which I invited a handful of philosophers to write on various technical features of contemporary economic theory and an equal number of economists to reply. (James Woodward and Nancy Cartwright contributed important essays on causal modeling in econometrics to this volume.) My own contribution was on the subject of the epistemic status of "computable general equilibrium models," a tool that was very much in vogue in development economics in the 1990s as a way of performing "experiments" involving changes in macro-economic variables and measuring the simulated results that came from these interventions. Eventually I published a book that resulted from the learning I did in these years, focused on the normative side of development theory, *The Paradox of Wealth and Poverty: Mapping the Ethical Dilemmas of Global Development*.

Which social sciences do you consider particularly interesting or challenging from a philosophical point of view?

I currently find one component of contemporary sociology most interesting, the field of comparative historical sociology. This is the body of work that involves comparative historical analysis of social institutions, processes, identities, or outcomes. Why do social revolutions occur or fail? Why are social welfare regimes so different across Western Europe and North America? What explains the different levels of militancy of East Coast and West Coast dockworkers? Researchers in this field are interested in discovering concrete social causes of important processes, and they endeavor to do so through somewhat detailed study of comparable historical cases. This is "small-N" research. Especially important leaders in this field were Charles Tilly, Theda Skocpol, and Barrington Moore, Jr. ; second – or third – wave contributors include Jack Goldstone, R. Bin Wong, George Steinmetz, and Julia Adams. Very important current work in the field includes contributions by Andrew Abbott, Kathleen Thelen, and Paul Pierson. Comparative historical sociology is a body of work that encompasses some degree of interdisciplinarity, in that some political scientists and anthropologists have also contributed to the literature. This approach has much in common with another particularly fruitful strand of research in sociology and political science, the "new institutionalism", since researchers in this tradition are particularly interested in discovering the differences in outcomes that are created by seemingly minor differences in the design of an institution.

This area of sociological research is particularly important, in my mind, for several reasons. First, these authors have largely rejected naturalistic models for social-science knowledge. They are not expecting to find exceptionless, cross-context "laws of society." Instead, they emphasize "constrained contingency." The ideas of path dependence, conjuncture of causes, contingency, and multiple possible causes and outcomes are embedded in the sociological imagination among these researchers. Second, they are nonetheless committed to finding explanations of social outcomes, and this means finding causes and constraints that lead collective behavior in one direction rather than another. For example, why did the Iranian Revolution take a very particular course of development, distinct from that of China and Cuba? Researchers have examined some processes in the three cases that are in common (social contention, social mobilization, social organization) and some features that are importantly different, and have come to analyses of the three revolutions that explains various features of their trajectories without hoping to reduce them all to a single theory of revolution. Third, these researchers emphasize rigor and methodological clarity, but they recognize the shortcomings of an exclusively statistical and "measurement of variables" methodology for the social sciences (large-N studies). They do not attempt to reduce the cases they consider to a small set of variables to be coded. Fourth, all these researchers have a common conviction that history matters; that the circumstances that were on the ground at the time of an important social change were themselves the result of important historical conditioning, and that it is an important piece of the sociological investigation to identify some of that historical setting. Finally, these researchers have often gone a bit further than some other areas of social science, in recognizing that "culture" matters, and that the tools of ethnographic interpretation are relevant and productive in areas outside of anthropological fieldwork.

Comparative historical sociology and the new institutionalism are significant examples of innovative social science for two reasons. First, the results are powerful, surprising, and insightful; this field of research has born very significant fruit. And second, the success of the field appears to have much to do with the cluster of background assumptions that researchers bring about the nature of the social world: for example, the contingency of social processes, the plasticity of social institutions, the malleability of human behavior, and the historical constructed character of social identities. These background ontological assumptions allow comparative historical sociologists to construct research methods that are indeed appropriate to the subject matter.

How do you conceive of the relation between the social sciences and the natural sciences?

The social world is, of course, embedded within the natural world, and human beings are themselves natural organisms. Moreover, there are features of the natural order that are directly pertinent to social explanation: cycles of climate change, the hydrology of great river systems over a time frame of centuries, the mechanisms of proliferation of disease. And features of human cognition and memory—the proper subject matter of a natural-science precinct of psychology—are plainly pertinent to the explanation of social phenomena (the observed size limitations on voluntary forms of cooperation, for example). So we do not need to draw a bright line between the scope of the natural sciences and the distinctive subject matter of the social sciences. All that said, I believe it is very important to recognize that systems of social phenomena are highly dissimilar from systems of natural phenomena. Social entities have their properties in virtue of the behavior and dispositions of the individuals who make them up at a given time—unlike complexes of molecules or strata of soils, whose constituents have uniform and timeless characteristics. Human action is contingent, motivated, self-modifying, and plastic. And therefore social institutions and processes lack determinate and fixed properties. Herbert Simon's "science of the artificial" provides a better metaphor for the social sciences than does the idea of the science of the natural world; social institutions and constructs are more analogous to systems of technology and artifact than they are to causally ordered natural systems. Chemistry, physics, and biology do not provide useful metaphors or models for the representation and explanation of social systems. "States" are not like "metals," with a common underlying causal reality. Stanley Lieberson's studies of social phenomena that are plainly not law-governed but nonetheless sociological provide excellent models for how to think of the subject matter of sociology—professional sports, patterns of first names, or the ways in which styles change over time.

What is the most important contribution that philosophy has made to the social sciences?

Important contributions from philosophers to the understanding and conduct of social science research might include these: clarification of the foundations of rational choice theory and the theory of rationality; clarification of the nature and mechanisms of collective action and collective entities; analysis of the meaning and methods of causal inference; explication of the notion of "causal mechanisms"; elaboration of the theory of supervenience as a solution to the relationship between social facts and facts about biological individuals; critique of functionalism as an explanatory strategy; re-interpretation of Marx's economic theories along the lines of "rational-choice Marxism."

Some of the contributions of philosophy to the social sciences have been on the detrimental side of the ledger: positivism, behaviorism, naturalism, the unity of science, and deconstruction, to name several. The deficiencies of these master theories of the social sciences derive from a common failing: the idea that there ought to be a single philosophical perspective that will drive the organization and development of social science knowledge. There is a moral here: when philosophers undertake to offer prescriptions for the social sciences at the highest level of generality, they generally miss the mark. When they focus on more specific issues that are of real working concern to social scientists, they make a meaningful and forward-moving contribution.

It is tempting to believe that continental philosophy may have made more of an enduring contribution to social inquiry through the elaboration of the ontology and methods of hermeneutics. It is true that there is a closer relationship between the "human sciences" and continental philosophy than between contemporary sociology and analytic philosophy. But much of that influence probably proceeds from working anthropologists such as Clifford Geertz rather than from technical hermeneutic philosophy. Analytic philosophy has the potential for contributing to social science research through some of its characteristic strengths when philosophers turn their attention to the social sciences—clarity, logical analysis of conceptual problems, and an insistence on rational standards of justification. These methodological characteristics are most valuable when exercised in concert with a real knowledge of and sympathy for the current research practices of innovative social scientists. Harold Kincaid's work fits this ideal, as does that of Daniel Hausman in the philosophy of economics.

Which topics in the philosophy of social science will, and which should, receive more attention than in the past?

Topics that probably will receive more attention—perhaps with diminishing returns:

- The role of laws and generalizations in social science
- The degree to which social science theories do or should resemble theories in the natural sciences
- Further debates about individualism, holism, and reductionism
- Conceptual or ethical relativism as a putative discovery of anthropological study of radically different cultures
- Covering law model as an ideal for social-science explanations

Topics that will yield valuable results for the philosophy of social science:

- Social ontology—more focused investigation of the nature of social entities, ensembles, structures, organizations, and events
- Social causation—more detailed theorizing about the nature of social causation. How are social-causal powers conveyed?
- Exploration of agent-based modeling as a way of exploring the theoretical consequences of assumptions about motives, constraints, opportunities, and social institutions
- Critical studies of quantitative methodology—better analysis of the scope and limits of statistical reasoning about social entities, processes, and outcomes
- Exposition of the limits that exist on prediction and the use of social and behavioral research to produce "social-engineering" applications to social policy.

- Studies of the logical and historical relations that exist among the disciplines and their domains of inquiry.

What are the most fundamental observations about the social sciences that you have come to through your research?

I have come to a small handful of central and iconoclastic reflections about social science theory, research, and explanation that I think are particularly fundamental:

- The social world is heterogeneous in multiple ways: causally, institutionally, organizationally, historically, and behaviorally.
- Social "things" (organizations, institutions, structures) are plastic and modifiable, and they do not fall into "social kinds" (in analogy with natural kinds).
- Social change is contingent and multi-causal.
- Social causation works through the mechanism of individual agents within concrete social settings, and nothing else.
- We need to look for "microfoundations" of social processes, causes, and facts.
- The key to social process and change is the action of the socially-situated agent making deliberate choices within given circumstances.
- Social identities and psychologies are themselves the product of prior "microfoundational" processes.
- We should not expect strong unifying theories that will "explain" all or most social phenomena.
- We should not expect strong regularities or laws among social phenomena.
- The natural sciences are a poor model for conceptualizing the nature of the social sciences.
- The social sciences are currently in need of some very basic new thinking, and philosophy of social science can play a very helpful role in this re-thinking.

9

Steven Lukes

Professor of Sociology
New York University, USA

How did you get interested in the philosophical aspects of the social sciences?

I was corrupted from the start. I studied PPE (philosophy, politics and economics) as an undergraduate at Oxford. At Balliol College, where I studied, philosophy was what gave the whole course intellectual coherence—perhaps because you could still study all three to quite a high level without much technical expertise, and thus spend time reflecting on conceptual issues and basic presuppositions. One of my economics tutors, Paul Streeten, was intensely interested in questions about values and objectivity in economic theorizing (and the writings on these themes by the Swedish sociologist-economist Gunnar Myrdal, which he edited [Myrdal 1958]). The other economics tutor, Thomas Balogh, was consumed by hatred of neo-classical economics and so challenged us to explain what was wrong with the orthodox textbooks (notably Paul Samuelson's *Economics*) and other mainstream writings that he made us read. In politics we read American political scientists and democratic theorists writing, in celebratory vein, about 'pluralist' democracy in the United States. Their supposedly objective findings and conclusions seemed to betray a distinct ideological complacency which, I already suspected, stemmed from conceptual and methodological choices—a suspicion I later elaborated in my *Power: a Radical View* (Lukes 1974).My first philosophy tutor (at the age of seventeen) was Charles Taylor, who was soon to write his first book *The Explanation of Behaviour* (Taylor 1964), criticizing behaviorist psychology. Another major presence was Sir Isaiah Berlin, Professor of Social and Political Theory. One of his compelling themes was the celebration of imaginative empathy in the interpretation of cultures in thinkers such as Vico and Herder and an implicit critique of scientism and 'positivist' social science. (He always thus regarded sociology). Another, of course, was his critique of theories of inevitability in philosophies of history.

As I moved into graduate work in the early 1960s, sociology was livening up and in the air, even the Oxford air, though I went to the LSE to Tom Bottomore's seminars to learn about it. I soon decided to write a doctoral thesis on Emile Durkheim, which I did under the supervision of the great social anthropologist E. E. Evans-Pritchard. What decided me to do so was my excitement at reading Durkheim's *The Elementary Forms of the Religious Life*, in part because of its and his sociological slant on a whole range of recognizably philosophical questions, and not least its bold, if misconceived, attempt to offer a sociological alternative to Kant by advancing a sociological account of the fundamental categories of thought. Durkheim was, of course, himself a philosopher-turned-sociologist but the philosophy from which he turned lacked rigor and indispensable tools that twentieth-century philosophers and logicians were to bring. (John Searle has recently made this observation (Searle 2006), and on this point I agree with him. More about this below). Working through the corpus of Durkheim's writings I deployed analytical philosophical distinctions and arguments to clarify my understanding of his ideas and to display their conceptual architecture.

I also read the early writings of Ernest Gellner—*Words and Things* and *Thought and Change* (about which I published a shamefully unappreciative review) and I was intrigued by his sociologically challenging essay 'Concepts and Society' (Gellner 1962) about the role of contradictory ideas in social life. I liked his sardonic critique of the rather self-satisfied and complacent 'linguistic philosophy' of the time but knew that he seriously underrated its strengths and was always absurdly unappreciative of Wittgentstein. (Despite my review, we subsequently became warm friends and co-editors of the *European Journal of Sociology*, along with Bottomore, Raymond Aron and Ralf Dahrendorf). And I was drawn into the wide-ranging and still continuing so-called 'rationality debate' (Wilson 1970), about whether there could be alternative, culturally-based standards of rationality, a debate initiated by Peter Winch's book *The Idea of Social Science* (Winch 1958) and his remarkable article 'Understanding a Primitive Society' (Winch 1964). This interest developed partly through my increasing interaction with the Oxford social anthropologists but above all because of my friendship and intellectual collaboration with the wonderful Martin Hollis, whose pure and extreme rationalism provoked me, though we always shared an antipathy to relativism and a futile determination to put it to rest. Our engagement in the rationality debate led to co-teaching seminars (with Taylor and my philosophy of science colleague Bill Newton Smith) and eventually co-editing *Rationality and Relativism* (Hollis and Lukes 1982).And in those early years the inspirational Alasdair MacIntyre had arrived in Oxford and generously invited me, while I was still a mere graduate student, to co-teach a class on philosophy of the social sciences, at which I presented a paper, soon published in the *British Journal of Sociology*, on methodological individualism (Lukes 1966). By then it was already clear to me that there was a range of philosophical questions to which the work of social scientists, classical and contemporary, was of central significance, often putting them into a new light and raising new, often philosophical questions unasked by social scientifically innocent or ignorant philosophers. Yet, as I have indicated, I was no less impressed by the contribution analytic philosophy could make by rendering far more precise social scientific claims and ideas—as exemplified in my good friend G. A. Cohen's masterly *Karl Marx's Theory of History* (Cohen 1978), which surpassed all previous writings on historical materialism, offering extremely fine-grained accounts of functional explanation, of how one can distinguish between social and material facts, and of much else besides that is of both philosophical and sociological interest.

Which social sciences do you consider particularly interesting or challenging from a philosophical point of view?

My difficulty in answering this question arises from being unable to identify 'a philosophical point of view.' If there is such a point of view, I doubt that I possess it. What I do think is that one can view all the social sciences philosophically, that is, in a philosophical way. What does that mean? Perhaps that one can probe their presuppositions, elicit their methodological commitments, identify their 'priors' and so forth. One can analyze the complex causal structures that their explanations posit, identifying which are the relevant counterfactuals that they require, distinguishing spurious from explanatory correlations, and the like. But these, surely, are all tasks that are themselves internal to the social scientific enterprise itself: what theoretically minded and skilled social scientists should be doing themselves anyway. So what kinds of questions, if any, are distinctly philosophical questions?

Negatively, perhaps we can say that they are questions the pursuit of which goes beyond or bypasses the explication or defense of some particular explanatory approach (which may well involve the criticism of alternative approaches). Positively, perhaps we can say that they are questions that raise some perennial puzzle that needs to be solved by practicing social scientists and that any given such approach will take to be solved in one or another way. Here are some examples of such puzzles. How are we to understand collective agency? What distinguishes social facts? What constitutes the meaningfulness of human action? Are their non-physical laws? Are the best or proper explanations reductionist? How are we to make sense of contingency in social life? Are there limits to our knowledge of social and economic processes? Is all knowledge local knowledge? In what does human well-being consist? What is it to be rational? Are there alternative criteria of rationality? And so on.

I suggest that those who assume one or another answer to questions such as these, in elaborating or defending some explanatory approach stand on this side of the social science/philosophy boundary (if there has to be one), while those who pursue such questions because they find them intrinsically interesting and important stand on the other side. Some specific examples will perhaps make my point. On this side of the line I would place, for instance, Jon Elster's defense of methodological individualism, his critique of functionalist explanation and his advocacy of explanation in terms of social mechanisms, James Coleman's *Foundations of Social Theory* (Coleman 1989), which sets out, with numerous examples, ways in which rational choice models can explain social phenomena and Clifford Geertz's elaboration of what is involved in 'thick description' (Geertz 1973) and his defense of 'anti-anti relativism' (Geertz 1984). On the other side I would place John Searle's account of social institutional facts, as epistemically objective while containing ontologically subjective elements essential to their existence, Peter Winch's *The Idea of Social Science* (Winch 1958), which offers an account of meaningful behavior that sharply distinguishes social from natural science, and Amartya Sen's explorations of alternative conceptions of rationality and freedom and his account of what constitutes well-being in terms of functionings and capabilities. Thinkers of the former kind are concerned to advance and defend a particular research program to which they are committed. They are, typically, impatient of and bored, even irritated by questions such as those above. To thinkers of the latter kind such questions are meat and drink. (Some, a very few, such as Sen, are equally at home on both sides of the dividing line).Thus Searle is concerned with social ontology because he aims to identify, as a philosophical enterprise, the distinguishing features of human social reality in order to show how this fits into the one world that consists entirely of physical particles in fields of force and this leads him to reject theories which postulate further realities. Winch sought to develop what he took to be a Wittgensteinian approach to the understanding of social life by focusing on what is involved in 'following a rule.' And Sen has explored the questions indicated by advancing ideas that offer a more realistic way of thinking about and dealing with poverty and inequality. And so he challenges prevailing ethical theories, such as utilitarianism, notably the assumption that individuals make rational utilitarian choices about their own welfare, and the Rawlsian theory of justice, notably its misdirected focus on 'primary goods.'

So I answer that all the social sciences raise interesting, challenging and problematic philosophical issues and that I find those just cited particularly so, and particular the last.

How do you conceive of the relation between the social sciences and the natural sciences?

I don't think I have anything new to add to the old debate about the relation between the social and the natural sciences that has rumbled on ever since August Comte invented Positivism, and so I shall largely cite with approval the arguments of others with whom I agree. I do think that all the interesting positions lie in the central region of the continuum that stretches between the extremes of undifferentiated unity, on the one hand, and radical dualism, on the other—that is within the range of views that combine, in different ways, what is common as *science* with what is divergent because *social*. In my view, the former must include the commitment to seeking rigorous epistemic objectivity and the latter must include recognition of ontological subjectivity accessible in its distinctly human form through mutually accessible languages (alongside the impossibility of exact replicability and of isolating closed systems through experiment).

In social science the pursuit of epistemic objectivity faces distinctive challenges of which I will here single out one that interests me. I refer to what is often, with a nod to Weber, called the issue of 'value-relevance' in social scientific explanation, namely the role of what Weber called *Kulturwertideen* in the framing of the questions social scientists ask. Myrdal put it this way: 'There is no way of studying social reality other than from the viewpoint of human ideas... The value connotation of our main concepts represent our interest in a matter, gives direction to our thought and significance to our inferences' (Myrdal 1958: 1). I would rather avoid the term 'values' (which never helps to clarify one's meaning) and say that practical, social, political and moral concerns typically and unavoidably enter into what we want social scientific inquiries to explain. But how far or deeply do they enter? Do they merely guide or govern the selection of the *explanandum* or do they also influence the *explanans*?

Consider the concept of power. This is obviously a causal concept but in social contexts to have power is have the ability to bring about *significant* outcomes, and to exercise power is to bring them about. Of course one can have the power to bring about trivial effects, but when we talk about power and power relations and when, as social scientists, we study it, seek to locate it, assess it impact, even try to measure it, and when we compare the power of some with that of others we make implicit or explicit judgments about what is significant. My power will be greater than yours if the interests of those I can affect are more significant. (A judge able to impose the death sentence has greater power than one who does not). Moreover, we are interested in power for extra-scientific reasons—to assign various kinds of responsibility, for example, or to evaluate the functioning of social arrangements. And this means that we will select out from the complex causal processes at work the chains and links which are relevant in the light of those extra-scientific interests. Of course, we can always try to expunge the vocabulary of power and its cognates (coercion, authority, manipulation, domination, hegemony and so on) from our *explanatia* and replace it with technical terms or mathematical modeling,but will the answers then respond to the interests that motivate them?

As for the recognition of ontological subjectivity, this does indeed imply, as Charles Taylor has commented, that there is an unavoidably 'hermeneutical' component in the sciences of man, for man is 'a self-defining animal' (Taylor 1971: 55). He makes this comment at the beginning of what is, in my view, still the best single-essay treatment of this topic: 'Interpretation and the

Sciences of Man.' (Taylor 1971) The most striking corollary Taylor derives from this observation is that there is a radical difference between the natural and the social sciences—between metereology, say, and sociology–with respect to their powers of prediction. With changes in self-definition 'go changes in what man is, such that he has to be understood in different terms. But the conceptual mutations in human history can and frequently do produce conceptual webs which are incommensurable, that is, where the terms cannot be defined in terms of a common stratum of expressions' (Taylor 1971: 55). I like and agree with the quasi-Popperian conclusion that Taylor draws: that 'conceptual innovation ... in turn alters human reality', and so the 'very terms in which the future will have to be characterized if we are to understand it properly are not all available to us at present.' (Taylor 1971: 56). It is, perhaps, worth noting that Taylor's arguments were principally directed against one particularly powerful trend of the time, influenced by the lure of a prevailing natural science model of explanation, namely behaviorism.

Another philosopher who has reflected upon ontological subjectivity is John Searle, for whom what is distinctive of human social reality is that it consists in 'institutions' (such as money and property), which, according to Searle, are created and sustained entirely in individual minds by collective intentionality. They are created, through the constitutive power of language, by the assigning of deontic powers—rights, duties, obligations, responsibilities and so on—which in turn create desire-independent reasons for action. Thus Searle concludes that institutional social facts are epistemically objective but 'only exist in virtue of collective acceptance or recognition or acknowledgement.' (Searle 2006: 13). I shall say more about this account under the next question, but here it is worth noting that part of Searle's motivation (which I applaud) is to meet the challenge of sociobiology and its 'implicit message... that human beings are not different from other social animals and that the terms in which we need to understand human social behavior are essentially biological and above all evolutionary.' (Searle 2006a: 40-41). On the contrary, human social behavior is done for reasons, under the constraints of rationality and within human institutions it is done with the presupposition of free-will and empowered by humanly created deontologies. This is a further deep and far-reaching way of distinguishing the social from the natural sciences.

9. Steven Lukes

What is the most important contribution that philosophy has made to the social sciences?

Philosophical contributions to social science can be direct when reflection on philosophical issues, in the sense defined above, affects the practice of some discipline. Consider the immense impact of utilitarianism on the agenda and conceptual apparatus of neoclassical economics. It has had a comparable influence upon the considerable swathe of political science that has been colonized by rational choice theory. More specifically, individual contributions can deploy their philosophical reflections to explanatory and, indeed practical effect. Consider, as already mentioned, Sen's parallel but interrelated critique of welfarist utilitarianism and his major contributions to welfare and development economics and Hayek's deployment of his arguments concerning our necessarily limited access to the relevant knowledge to the winning of the socialist calculation debate and his assault on the very idea of central planning. Such contributions can, of course, be 'important' but deleterious and social science can be influenced by poorly understood or bad philosophy. Talcott Parsons's critique of what he took utilitarianism to be in developing his 'theory of action' exemplifies the former; and the adoption of extreme relativist and post-modernist ideas in some areas of recent 'interpretivist' cultural anthropology exemplifies the latter. I suppose that bad or poorly understood philosophy could generate good social science. (Perhaps the so-called 'strong program in the sociology of science is an instance of this). Do good social scientists (or good scientists in general) need to be good or even decent philosophers, or indeed even philosophers at all? It is certainly not obvious that the normal science of progressive research programs benefits by focusing attention on their presuppositions.

This leads me to consider a second, indirect way in which philosophy can contribute to social science: that is, precisely by focusing on its ontological and methodological presuppositions. Here, as just suggested, the effects on social science practice are not necessarily beneficial. The classical sociologists—notably Marx, Weber and Durkheim—wrote well-known methodological texts that have become canonical within the discipline. They are cited in textbooks and assigned to beginning students, yet they cry out for close analysis and rigorous critique conducted with the benefit of advances in logic and philosophy that have occurred since they were written. I have already cited Cohen's discussion of Marx's historical materialism, which itself generated much debate and further treatments of a similar sort. Weber's account of *Verstehen* and causality have benefited insufficiently from philosophically informed discussions of these issues. Much could doubtless be gained by drawing in the former case on the 'rule-following debate' deriving from Saul Kripke's reading of Wittgenstein, and in the latter on contemporary discussions of causation.

As for Durkheim, his *Rules of Sociological Method* and his other methodological writings constitute a perfect case for such philosophical treatment. As it happens, John Searle has mounted a spirited critique of Durkheim's views, but I maintain that in criticizing Durkheim Searle in effect reveals with great precision what is implicit and unworked out in his thinking. Searle comments that Durkheim 'was not in a position to state the precise details of collective representations because he lacked the logical apparatus. (This is not his fault. He happened to live in the prehistory of the study of intentionality).' (Searle 2006b: 62). Searle complains that Durkheim has a mistaken conception of society and social facts, failing to see that we can have an epistemically objective science of a domain that is ontologically subjective—that, as stated above, these objective facts only exist 'in virtue of collective acceptance or recognition or acknowledgement.' Searle holds that Durkheim did not grasp the distinctions needed to see this and the paradox it involves: that institutional social facts are real but only exist because people think they exist. I maintain (and have argued elsewhere [Lukes 2007]) that Durkheim's writings abundantly express, but do not analyze, this idea and paradox. Moreover, I further maintain that Durkheim, as a sociologist, addresses a problem which Searle, as a philosopher, leaves entirely unaddressed, namely, how is 'collective acceptance or recognition or acknowledgement' achieved? Durkheim, in short, begins where Searle leaves off by inquiring into the transmission and inculcation of ideas, from the most concrete and specific to the basic categories of thought, into the mechanisms of socialization, into the conditions under which the recognition of shared norms and integration into shared values break down and the consequences of such breakdown, and into the role of ritual and symbolism in generating and regenerating enthusiastic support for the collective representations that unite communities.

Which topics in the philosophy of social science will, and which should, receive more attention than in the past?

My answer to this is to point to a domain that used to be jointly occupied by social scientists and philosophers but has been vacant for some five decades. We can label this domain 'the sociology and psychology of morals.' It was a central topic at the turn of the twentieth century chiefly among philosophers, sociologists and anthropologists (the distinction between the latter was not yet institutionalized). In France Lucien Lévy-Bruhl and Durkheim, in Britain L. T. Hobhouse and Edward Westermarck, in the USA Franz Boas and William Graham Sumner and later Margaret Mead and Ruth Benedict all operated within this territory. Moral philosophers ventured into it in the 1950s, notably John Ladd (Ladd 1957) and Richard Brandt (Brandt 1954). I believe that much could be gained at the present time from a convergence between sociologists and anthropologists, on the one hand, and moral philosophers, on the other (see, for example Moody-Adams 1997, Cook 1999, Moser and Carson 2001 and Levy 2002). Moreover, there is now an accumulating body of work by psychologists, cognitive scientists and experimental or behavioral economists, working in collaboration with philosophers, exploring how individuals acquire, implement and are affected by norms, the possible relations between innate mechanisms and cultural learning in the acquisition of moral norms and, in general, the psychology of moral norms (Carruthers et al. 2006 and Hauser 2006).

There are some fascinating questions to be explored here, questions that have been insufficiently explored at the appropriate level of philosophical sophistication. What sense can be made of the claim that there is 'a diversity of morals'? How are we to distinguish between moral principles, rules, norms, codes and practices? How do sociological and anthropological accounts of what morality is differ from philosophical accounts? What can the latter learn from the former and the former from the latter? What evidence can be adduced for the diversity of morals within and across cultures? If it exists, how wide and how deep does such diversity go and how can the answer to these questions be established empirically? Is 'morality' an analytically useful concept that can be applied cross-culturally? Can we reconcile its external use, by observers engaged in description and explanation, and its internal or first-person use, by moral agents or participants. Are their distinguishable domains of morality that can be identified across cultures? Is there a convincing (non-relative) way to distinguish between what is customary (*mores*) and what is moral? What is the force of the so-called enculturation thesis in accounting for how people acquire moral principles, beliefs and attitudes? What is the value of the evidence gathered by psychologists of subjects' responses to moral dilemmas, generated by philosophers, such as the so-called 'trolley problem' purporting to show the ways in which they display the capacity for moral reasoning and judgment? Is that capacity universal or species-wide among humans? What is the bearing of studies of animals on answering this question? How strong is 'the linguistic analogy' drawn from Chomskian linguistics with morality—an idea mooted at one point by John Rawls? Can we plausibly postulate a universal moral grammar, with unconscious principles and parameters? (Hauser 2006 and Susan Dwyer in Carruthers et al. 2006). If so, what would be the scope for reasoning, deliberation and reflection (in which we typically seek the 'right answer' to moral questions and dilemmas) and what would this imply for the thesis of moral relativism? (After all, the principles and parameters of Chomskian universal grammar result in our speaking one language or another, or several: there is no Esperanto-like location from which we could arrive at the 'right' way to speak). How does moral relativism differ from other forms of relativism, and how strong are the arguments for and against it? Which human rights should be universal? Some of these questions are sociological, some psychological, some are philosophical and most are inextricably all at once.

References

Carruthers, P., Laurence, S. and Stitch, S. (2006) *The Innate Mind.* Volume 2: 'Culture and Cognition.' New York: Oxford University Press.

Cohen, G. A. (1978) *Karl Marx's Theory of History. A Defence.* Oxford: Clarendon

Coleman, J. (1989) *Foundations of Social Theory.* Cambridge, Mass: Harvard University Press

Cook, J. W. (1999) *Morality and Cultural Differences.* New York and Oxford: Oxford University Press

Geertz, C. (1973) *The Interpretation of Cultures.* New New York: Basic Books

Geertz, C. (1984) 'Anti Anti-Relativism', *The American Anthropologist.* 86(2): 263-78 reprinted in Geertz, C. *Available Light: Anthropological Reflections on Philosophical Topics.* Princeton: Princeton University Press, 2000

Gellner, E. (1962) 'Concepts and Society', *Transactions of the Fifth World Congress of Sociology* (Washington). Louvain: 1962, 1, pp. 153-83, reprinted in Gellner, E., *Cause and Meaning in the Social Sciences*. London and Boston: Routledge and Kegan Paul, 1973.

Hauser, M. D. (2006) *Moral Minds: How Nature Designed Our Universal Sense of Right and Wrong*. New York: Harper-Collins

Hollis, M. and Lukes, S. eds. (1982) *Rationality and Relativism*. Oxford: Blackwell

Levy, N. (2002) *Moral Relativism: A Short Introduction*. Oxford: Oneworld

Lukes, S. (1968) 'Methodological Individualism Reconsidered', *British Journal of Sociology*, 19, 2: 119-129

Lukes, S. (1974) *Power: A Radical View* London: Macmillan. Republished in second edition with introduction and two new chapters London and New York: Palgrave, 2004

Lukes, S. (2007) 'Searle versus Durkheim' chapter in Savas L. Tsohatzidis (ed.) *Intentional Acts and Institutional Facts: Essays on John Searle's Social Ontology*. Dordrecht: Springer

Moody-Adams, M. M. (1997) *Fieldwork in Familiar Places: Morality, Culture and Philosophy*. Cambridge, Mass.: Harvard University Press

Moser, P. K. and Carson, T. L. eds. (2001) *Moral Relativism: A Reader*. New York and Oxford: Oxford University Press

Myrdal, G. (1958) *Value in Social Theory: A Selection of Essays on Methodology*. Edited by Paul Streeten. London: Routledge and Kegan Paul

Searle, J. (2006) 'Social Ontology: Some Basic Principles' *Anthropological Theory*, 6(1) March: 12-29

Searle, J. (2006a) 'Culture and Fusion: Reply to D'Andrade,' *Anthropological Theory*, 6(1) March: 40-44

Searle, J. (2006b) 'Searle versus Durkheim and the waves of thought,' *Anthropological Theory*. 6(1) March: 57-69

Taylor, C. (1964) *The Explanation of Behaviour*. London: Routledge and Kegan Paul

Taylor, C. (1971) 'Interpretation and the Sciences of Man', *Review of Metaphysics*, 25: 3-51. also published in C. Taylor, *Philosophy and the Human Sciences: Philosophical Papers 2,* Cambridge: CUP, 1985, Ch 1 (page references to this version)

Wilson, B. R. (1970) *Rationality.* Oxford: Blackwell

Winch, P. (1958) *The Idea of a Social Science and its Relation to Philosophy.* London: Routledge and Kegan Paul

Winch, P. (1964) 'Understanding a Primitive Society,' *American Philosophical Quarterly,* 1: 307-24 reprinted in Wilson 1970)

10

David Papineau

Professor of Philosophy
King's College London, UK

How did you get interested in the philosophical aspects of the social sciences?

My first academic appointment was to a Lectureship in the Department of Sociology at the University of Reading in 1973. I suppose that this was a slightly odd appointment, given than my academic training until then had been entirely in mathematics and philosophy. But the Head of the Reading Department at that time was Stanislav Andreski, an émigré trained in the pre-war Polish logical empiricist tradition of comparative sociology. He thought that the undergraduates needed a solid grounding in basic logic and Mill's methods of induction and that I was the person to give it to them.

I was there for four years and it was a wonderful education for me. Anybody who has read Andreski's delightfully irreverent *Social Sciences as Sorcery* (1972) will know that he had no patience with sociological fashion. As a result, few of his appointments were conventional sociology PhDs. Instead he amassed a distinguished collection of intellectual waifs and strays. Many were from Europe, displaced by the war and its aftermath, and shunned by more conventional employers. The ethos of the department was empirical but by no means narrow. Bill Russell, who taught ethology and cultural evolution, had trained as a psychoanalyst. Viola Klein had been a student of Mannheim's and was one of the first sociologists to write on the role of women. Alexander Lopasic was an anthropologist from the then Yugoslavia, Maria Hercowitz was an industrial sociologist from Poland, and Peter (Tank) Waddington was a criminologist who had recently left the police force. Overall the focus of the Department was on the broad sweep of history, and it would be hard to imagine a collection of people who knew more about it.

My first book, *For Science in the Social Sciences* (1978), was based on my 'Methodology' lectures in Reading. In fact I didn't do much methodology in the lectures. Mostly I discussed the then current topics in the philosophy of social sciences: holism versus individualism, explanation versus understanding, historical determinism, traditional belief systems, fact and value. A few sections of the book now strike me as a bit dated, concerned with debates that have since gone dead, but for the most part I still stand by what I said then. As the title indicates, the book was an unabashed defence of the 'positivist' view that there is no great divide between the social and natural sciences. But I didn't say anything too extreme. Of course I denied that there is anything magical about humans or societies that puts them beyond the limits of normal empirical investigation. But beyond that most of the book consists of sensible attempts to unravel theoretical tangles and avoid seductive fallacies.

Curiously, the part of the book that I now like least is the section on general epistemology of science. My Head of Department had hired me to teach logic and Mill's methods, but mostly I'd been telling the undergraduates about Kuhn and Feyerabend. Not that I endorsed Kuhn and Feyerabend's relativism. I saw myself as upholding realism against their ingenious attacks on scientific objectivity. Except that it wasn't real realism that I was defending, but the sophisticated neo-Popperianism of Imre Lakatos. At that time philosophy of science in Britain had been hijacked by the Popperians, and we were all brought up to think that falsificationism was enough to make the world safe for science. Nobody pointed out that Popper and his followers weren't even close to realists, but simply out-and-out skeptics who insisted that it is always wrong to believe any scientific theory. I suppose it is understandable that I went along with this tosh—plenty of more distinguished figures than I were equally tainted—but I can't help feeling bad about it now.

Which social sciences do you consider particularly interesting or challenging from a philosophical point of view?

I'm disinclined to pick out any one particular social science as particularly philosophically challenging, if only because I find it difficult to take the institutional divisions between the different social sciences seriously in the first place. It has always seemed to me that the subject matters of sociology, anthropology, history, political science, social psychology and even economics are so intertwined as to make the traditional divisions between the disciplines seem artificial. I know that the different subjects have different institutional traditions, with different canonical figures and different ideas on how young scholars should be trained. But in my view the disciplinary introversion that this encourages is a vice, not a virtue. The best work in all areas is typically informed by thinking from other branches of social science.

One question worth asking about the social sciences taken collectively is—'what are they trying to find out?' Back in the 1970s I thought I knew the answer: the social sciences, like all empirical sciences, aim to uncover general laws. But now this seems to me far too narrow a conception of what the social sciences can achieve at a theoretical level. Not that I think that there are no serious laws in the social sciences. There are plenty of striking patterns supported by the empirical data, both at a large-scale and individual level. (If you are skeptical, here are two empirical hypotheses that have become prominent since the 1970s: there are no wars between democracies; there are no famines in democracies. Or, to switch to the individual level, consider the findings of Milgram's obedience experiment, or of ultimatum game experiments.) Still, it now seems to me quite misguided to hold, as I used to, that all interesting theoretical findings in the social sciences can be fitted into the mould of general law.

A first point here is one that applies to the natural as well as the social sciences. We can often know about the causes and mechanisms that operate in certain local contexts without knowing any laws that cover them. This is something that Nancy Cartwright has been stressing for years (1983, 1989, 2000). Causes typically come before laws, at least in the order of discovery. Cartwright originally stressed the point in connection with the physical sciences. But the moral applies all the more strongly in the social realm. Social investigations uncover huge numbers of causal truths, but very few laws.

Of course, we can agree with Cartwright on this methodological point without endorsing the heavy-duty metaphysical consequences that she takes to follow. I myself remain an unrepentant Humean about causes, at least to the extent of thinking that causes can be reduced to non-causal laws, and in particular to probabilistic laws that impose the requisite asymmetry on causal relationships (cf Papineau 1991). But I am happy to agree that, at a methodological level, the Humean constitution of causes is often epistemologically irrelevant. We typically pinpoint causes using techniques that don't require any identification of covering laws.

Randomized experiments make the point graphic. They can tell us that some treatment affects some result, but not how the size of this effect will vary with circumstances. The same point applies to 'Bayesian Net' methods for identifying causes from survey data. Indeed we can usefully view both these techniques as probabilistic analogues of even simpler forms of everyday reasoning codified as Mill's 'methods of induction'. To take the very simplest case, if some occurrence of interest is determined, and we know that only one change preceded it, then we can be sure that this was the operative cause in the circumstances, even if we don't know what other factors were required for it to produce the effect. Randomized experiments and survey analyses simply apply this everyday form of reasoning to probabilistic data.

As I said, these points about knowledge of causes and mechanisms apply to the natural as well as the social sciences. However, there is one specific kind of mechanism that does distinguish the social from the natural sciences. Many social processes are driven by mechanisms involving intentions. What happens depends on what people believe and desire.

I don't think that the structure of intentional mechanisms is nearly as well understood as people generally suppose. Back in the 1970s, along with pretty much everybody else, I was an unreconstructed internalist about intentional explanation. Intentions were internal states that explained actions. We used to argue about the status of these explanantions, and in particular about whether they yielded some distinctive kind of insight that was not on offer in the natural sciences. But even those who were 'anti-positivist' about such explanations still thought of them as explaining bodily actions, things you did in your own space, so to speak.

This internalist perspective now looks far too limited. Over the last few decades philosophers of mind have become persuaded that the contents of mental states are externally constituted. In principle, two thinkers could be molecule-for-molecule identical, yet their beliefs and desires refer to different things, because of differences in their environment and background. (Think of the same mental symbol having different meanings in the two cases, because it is being used to pick out different things—my physical duplicate is thinking of *his* wife, I'm thinking of *my* wife.)

Philosophers of psychology, especially Jerry Fodor, haven't liked this. We use intentional states to explain bodily behaviour, they urge, and externally constituted intentional states are no good for this. How you move depends on what is going on inside you. External differences that don't correspond to internal differences can't make a difference. (Fodor 1980, Segal 2000.)

I agree with Fodor that externally constituted states are no good for explaining bodily actions. But I don't think that this means that they are no good for explaining anything. The moral is that intentional explanations explain things in addition to bodily actions—namely, interactions between agents and their environments. The mistake is to suppose that intentional psychological explanation stops at the bodily peripheries. Rather, intentional explanation sees agency as extending out into the environment, and gives us a grip on how this works.

On reflection, it is scarcely surprising that intentional explanations should operate like this. After all, intentional states are individuated by truth or satisfaction conditions—that is, by specifications of how things have to be in the external world for beliefs to be true or desires to be satisfied. However, this representational dimension structure would surely be quite otiose if the only explanatory purpose of intentional states were to account internally for bodily behaviour. There's no obvious reason why the internal states that interact to cause bodily behaviour should also represent features of the external world. They would surely play their behaviour-generating role just as well even if they were 'blind syntactic' states whole nature was exhausted by their internal causal powers. (Cf Papineau 1993 ch 3.)

I would say that the explanatory point of ascribing intentional states is precisely to chart the impact agents will have on their environments when things go as they plan. From this perspective, there's nothing puzzling about the fact that intentional contents are often externally constituted—that's just what we should expect if we use them to track the way agents interact with their environments.

I'm not sure how far this issue is appreciated by philosophers of social science. The case for externalism about mental states has largely been made by analytic philosophers of mind, and has rested on intuitions about possible cases—twin-earth water, thigh-bone arthritis, swampman (Putnam 1975, Burge 1979, Davidson 1987). However, the analytic philosophers of mind have paid relatively little attention to the special explanatory resources offered by externally constituted contents. This is one reason why internalists like Fodor have been able to argue that content externalism is an artefact of everyday intuitions and signifies nothing of explanatory importance. More work needs to be done in understanding the way intentional contents provide a bridge which relates the activities of agents to the wider world around them.

There is a great deal of interest among contemporary philosophers of social science in collective agency: do some supra-individual wholes qualify as agents in their own right? This is an important issue, but I would say that a prior task is to understand the structure of agency per se. Whether we are dealing with individual or collective agents, what exactly is that gets explained we explain their activities in intentional terms? It seem to me that we will have a much better understanding of the intentional mechanisms that are peculiar to the social sciences when we can answer this question.

How do you conceive of the relation between the social sciences and the natural sciences?

As I said earlier, I don't think that there is any great divide between the social and the natural sciences. People and societies are both parts of the natural world, and should be studied as such. Of course, there are particular investigative methods that are appropriate to particular social phenomena, just as there are particular methods appropriate to geology, hydrodynamics and plate tectonics. But this just shows that special kinds of complexity call for special investigative techniques, not that these different systems are constituted out of different fundamental ingredients.

On the question of fundamental ontological constitution, I am not just a naturalist but more specifically a physicalist. I think that everything is physically constituted. To invoke Saul Kripke's illuminating metaphor, once God had created all the physical stuff, his work was done—he had already made the people, thoughts, football clubs and nation states, by putting all the molecules in the right places.

Not that I take physicalism to be an a priori thesis. There is nothing contradictory in the idea that reality contains various kinds of non-physical stuff. This was certainly the orthodox opinion among mainstream scientists until little more than a century ago. But I take it that twentieth-century science, and in particular modern physiological research, has given us strong reason to doubt the existence of any non-physical entities. (Cf Papineau 2002 Appendix.)

Still, even if everything is physical, including the social realm, it doesn't follow that there are any interesting *epistemological* relationships between physics and the social sciences. In particular, it certainly doesn't follow that any claims made about the social realm need to be shown somehow to follow from the laws of physics. Non-fundamental natural sciences again provide a model. I take it that the subject matter of plate tectonics is physically constituted, if anything is. Yet this doesn't mean that the principles of plate tectonics need to be legitimated via some reduction to physics. Rather, they owe their acceptance directly to seismological, geological and other empirical evidence. Similarly, even if people and societies are physically constituted, we can seek to understand their workings by direct empirical means, without worrying about their physical underpinnings.

Is this a practical or principled point? It might seem that, if a system is physically constituted, then *in principle* it must be possible to understand it in physical terms, even if in practice this proves impossibly complicated. Some dispute even this much. Jerry Fodor (1974) and others have argued that the categories of the special sciences are 'variably realized' at the physical level, and that this constitutes an in-principle barrier to any uniform reduction of special scientific patterns to physical ones. To repeat Fodor's celebrated example, what counts as *money* will be physically quite different in different societies, so there can't possibly be any uniform physical story to explain the principles of monetary economics.

I have never been comfortable with Fodor's strong anti-reductionism about the special sciences. If the world is at bottom physical, shouldn't we expect systems with different physical realizations to behave differently? Putting it the other way around, if we observe some regular special scientific pattern, doesn't that argue that the special categories in question must have some uniform physical basis? From this perspective, I would argue that Fodor's money example is spurious. Money might be physically different in different societies, but it is *psychologically* the same everywhere—people expect others to provide goods and services for money, and therefore desire it themselves. And if psychological categories are uniformly realized at the physical level, then this would restore the in-principle possibility of a physical reduction of economic principles.

This kind of argument for in-principle reducibility is controversial. Some argue that natural selective processes of various kinds can give rise to genuine variable realizability and so block even in-principle reducibility. The idea is that different physical mechanisms often play the same psychological or social role precisely because they have all been *selected* to play that role. This then renders it unsurprising that the relevant mechanisms should be physically disparate—natural selection doesn't care, so to speak, about which mechanism plays some functional role, as long as it does the job. (Macdonald 1992, Papineau 1993 ch 2, Block 1997.)

So maybe Fodor's in-principle anti-reductionism can be saved by an appeal to Darwinian processes (which would be ironic, given Fodor's well-known antipathy to Darwinian thinking of any kind). Still, if we do side with Fodor against the in-principle reducibility of the special sciences, it is important not to do so for the wrong reasons. Fodor sometimes writes as if in-principle reducibility would impugn the autonomy of the social sciences and mean that sociology departments ought to be taken over by physicists. But this does not follow at all. In-principle reducibility is a metaphysical thesis with no practical consequences. Even if the psychological and social realms were in principle reducible to physics, this wouldn't make any difference at all to practising social scientists. They would still investigate their subject by direct empirical means, with no concern for physical realizations.

By way of analogy, nobody thinks that molecules are variably realized at the physical level and that there is therefore an in-principle barrier to the reduction of chemistry to physics. But we still have chemistry departments as well as physics departments, for the familiar reason that it is impracticable to derive knowledge of chemical systems from our knowledge of the basic physical principles governing their components. The same point applies to sociology departments. We will still need them in practice, even if it is principle possible to reduce social truths to physics.

Which topics in the philosophy of social science will, and which should, receive more attention than in the past?

If there is an epistemological connection between the social and natural sciences, it involves biology rather than physics. Humans are biological beings as well as physical ones. While our physical nature may be largely irrelevant to our social being, the same is not necessarily true of our biology.

Not that social scientists have a very good record of thinking about this connection. Ever since the modern synthesis of Darwin and Mendel, the application of biological ideas to social phenomena has been blighted by a tendency to dichotomize human characteristics into the natural or the nurtured, the instinctual or the cultural, the innate or the acquired.

Of course, some improvements have been made. E.O. Wilson's 'sociobiology' (1975) corrected the naivety of the mid-twentieth-century assumption that any animal 'instincts' would automatically be for the good of the group. And the 'Evolutionary Psychology' movement of the 1990s corrected the naivety of sociobiological assumptions about the relation between contemporary behaviour and the past evolution of cognitive mechanisms. (Barkow, Cosmides and Tooby eds. 1992.)

But even the Evolutionary Psychologists tend to work with a crude and unexamined notion of 'innate' cognitive mechanisms. Part of the problem here is conceptual. What exactly is it for something to be 'innate'? Those who trade in this notion find it very hard to explain what they mean. 'Causally quite independent of environmental factors' is clearly far too strong—everything depends inter alia on environmental factor, including legs and arms. 'Unalterable by interventions' is also too strong, for just the same reasons. Perhaps the best that can be done, at least in connection with cognitive traits, is to equate 'innate' with 'normally develops without any help from learning or other psychological processes' (though this does leave us with the non-trivial task of explaining what 'normal' development and 'psychological' processes are). (Cf Griffiths 2002, Samuels 2002.)

But behind this conceptual awkwardness lies a more substantial issue. Are there really any innate cognitive traits? I myself think that our genetic heritage plays a huge role in shaping cognitive and hence social structure, but at the same time I doubt that *any* cognitive structures are 'innate', even in the weak sense of normally developing without learning. In my view, it's learning all the way down, right back to the earliest stages of ontogeny—but learning that is strongly biased from the start by genes that have been selected because they make us especially good at learning certain things.

There seem to me principled reasons for adopting this view. I've always wondered how the complex sets of genes supposedly determining Language Acquisition Devices, or Theories of Mind, or Cheater Detection Mechanisms, could have become fixed in our ancestral populations. There are well-known barriers to the fixation of such pluralities of genes, arising from their individual non-advantageousness. I can remember once airing this worry to my then student (and now colleague) Matteo Mameli. 'But what if proto-versions of these cognitive traits were originally due to general learning mechanisms?' he asked. 'Then you would expect a quick accumulation of genes that made people better at learning these traits.' I burst out laughing at the thought that all the prize exhibits in the nativist list of cognitive traits might be congealed versions of capacities that were originally due to general learning mechanisms. But of course Matteo was quite right. His suggestion happily explains why there is no evolutionary barrier to the selection of the genes that make us so very good at acquiring language, understanding other minds, and doing the other things that come so easily to us (Papineau 2005).

Note how this picture does justice to the 'poverty of the stimulus argument'—it explains why humans learn certain things with very minimal informational input—without requiring that any cognitive traits are 'innate' in the sense of requiring *no* learning. Indeed the picture gives us reason to doubt that there is ever innateness in this strong sense. After all, why should evolution bother to substitute genes for *all* informational input, so to speak, if that input is freely available in nearly all environments? (Babies could in principle have been genetically designed to 'grow' the ability to discriminate phonemes even if they never heard any human speech; but what would be the point, given how few suffer such impoverished environments?)

These considerations indicate that most of the important cognitive traits underlying human social organization are a product of *both* culture and genes. We have evolved certain genes because they made us good at acquiring certain intellectual traits; and these traits now come naturally to us because we have those genes. Of course, these genes also make us naturally good at other activities apart from those they were originally selected to foster (typing, mathematics, reading novels) and many of these are of great social importance. At the same time, the continued dependence of 'genetically natural' traits on learning means that they are by no means guaranteed to emerge in all modern environments.

The interdependence of genes and culture thus leaves us with many important questions. How do genes and culture interact in the course of individual ontogeny? How did they interact in the evolutionary history of hominid society? What limits do their interaction place on the space of possible societies? It is not clear how to answer these questions. We need more data (historical , comparative, developmental, and experimental) and more theories. But the first step is to abandon the outmoded assumption that cognitive traits can usefully be divided into those that are innate and those that are acquired.

References

Andreski, S. 1972. *Social Sciences as Sorcery* London: Andre Deutsch

Barkow, J. Cosmides, L. and Tooby, J. eds. 1992. *The Adapted Mind: Evolutionary Psychology and the Generation of Culture* New York: Oxford University Press

Block, N. (1997) 'Anti-Reductionism Slaps Back' in Tomberlin, J. ed. *Philosophical Perspectives* 11

Burge, T. 1979. 'Individualism and the Mental' *Midwest Studies in Philosophy* 4

Cartwright, N. 1983. *How the Laws of Physics Lie* Oxford: Oxford University Press

Cartwright, N. 1989. *Nature's Capacities and their Measurement* Oxford: Oxford University Press

Cartwright, N. 2000. *The Dappled World: A Study of the Boundaries of Science* Cambridge: Cambridge University Press UP, 2000

Davidson, D. 1987. 'Knowing One's Own Mind.' *Proceedings and Addresses of the American Philosophical Association* 60

Fodor, J. 1974. 'Special Sciences or: The Disunity of Science as a Working Hypothesis' *Synthese* 28

Fodor, J. 1980. 'Methodological Solipsism considered as a Research Strategy in Cognitive Science' *Behavioral and Brain Sciences* 3

Griffiths, P. 2002. 'What is Innateness?' *Monist* 85

Macdonald, G. 1992 'Reduction and Evolutionary Biology' in Charles, D. and Lennon, K. eds *Reduction, Explanation and Realism* Oxford: Oxford University Press

Papineau, D. 1978. *For Science in the Social Sciences* London: Macmillan

Papineau, D. 1991. 'Correlations and Causes: Review Article of Nancy Cartwright *Nature's Capacities and their Measurement*' *British Journal for the Philosophy of Science* 42

Papineau, D. 1993. *Philosophical Naturalism* Oxford: Blackwell

Papineau, D. 2002. *Thinking about Consciousness* Oxford: Oxford University Press

Papineau, D. 2005. 'The Cultural Origins of Cognituve Adaptations' in O'Hear, A. ed *Philosophy, Biology and Life* Cambridge: Cambridge University Press

Putnam, H. 1975. 'The meaning of 'meaning' in Gunderson, K. ed. *Language, Mind and Knowledge, Minnesota Studies in the Philosophy of Science VII* Minnesota: University of Minnesota Press

Samuels, R. 2002. 'Nativism in Cognitive Science' *Mind and Language* 17

Segal, G. 2000. *A Slim Book about Narrow Content* Cambridge, Mass: MIT Press

Wilson, E. 1975. *Sociobiology: The New Synthesis* Cambridge, Mass: Harvard University Press

11

Philip Pettit

L.S.Rockefeller University Professor of Politics and Human Values

Affiliate Professor of Philosophy

Princeton University, USA

How did you get interested in the philosophical aspects of the social sciences?

I got into philosophy originally through becoming possessed by the work of Jean Paul Sartre; I wrote a short dissertation on the notion of bad faith in his philosophical and literary work for an L.Ph degree, when I was still in my teens. Later I came to see Sartre as mistaken in his exceedingly atomistic picture of the human subject, especially in his earlier work, and having found that bias in Sartre, I began to find it everywhere. That was a recurrent theme in my lectures at University College, Dublin as a young, pre-doctoral lecturer in the later sixties. It ushered in a period in which I searched for an alternative, less atomistic picture of the mind, which would allow for the intuitive interdependencies between people in society.

One body of literature I explored was associated with French structuralism but I became disillusioned with this approach and was quite critical of it when I wrote a short book, *The Concept of Structuralism*, in the mid 1970's. While working on structuralism, which I did as a post-doctoral fellow at Cambridge University, I became more and more interested in Wittgenstein, and in what I took to be his very non-atomistic view of how rule-following, and so thinking itself, is possible. This interest matured in the course of working my way into more analytical traditions in the philosophy of mind and language. I did that work in collaboration with Graham Macdonald, my colleague at the University of Bradford, which led to the publication of our book on 'Semantics and Social Science' in 1981.

11. Philip Pettit

The appearance of Saul Kripke's book on Wittgenstein on rule-following in 1982 had a big impact on me and on my continuing obsession with working out a non-atomistic picture of the mind and exploring its implications. In 1983 I moved from Europe to Australia, fleeing Margaret Thatcher's England, and over the following decade I was able to benefit from that wonderfully robust philosophical culture in developing this idea. The line I argued on rule-following laid the basis for the non-atomistic theory of mind presented in a number of articles and in my book, 'The Common Mind', which appeared in 1993. In that book I developed this view of the human subject and used it to argue for an associated complex of approaches in the philosophy of social science and in political philosophy.

The key idea in that book is that while individualism is sound, atomism is not. Individualism argues rightly that whatever forces are operative in society, they do not undermine our conception of ourselves as imperfect but more or less autonomous subjects. Atomism is mistaken, so I maintained, in suggesting that all the central aspects of the human mind are fixed in place without anything more than causal dependence on social relationships. I argued that the capacity of humans to reason in a mutually accessible way presupposes a process of triangulation on common meanings in which more than one person has to be involved. No one can become capable of reasoning, then, without relying on others; and this, for more than contingent, causal reasons. Being able to reason without any history of reliance on others would be like being able to applaud with one hand.

Which social sciences do you consider particularly interesting or challenging from a philosophical point of view?

Inevitably perhaps, the issue that I currently find most interesting is that on which I am currently working. I am working on the issue because I believe it is very interesting, not the other way around. Or so I would certainly like to think.

Perhaps the best way of introducing this interest is by reference to a third doctrine in social ontology, on a par with individualism and atomism. Individualism holds that the individual human mind, as it is conceptualized in our folk psychology, is not deeply compromised by the existence of those social regularities charted or likely to be charted in the social sciences; this doctrine I accept. Atomism holds the quite distinct theory, which I reject, that individual human minds do not depend on one another, except in an obvious causal manner, for the display of any distinctive characteristics. But there is a third doctrine that bears on a further issue about mind and society that I describe, in a term of Margaret Gilbert's, as 'singularism'. This holds that apart from individual minds, there are no other minds to be found in human society; in particular, group agents, incorporated collections of people, do not have minds of their own.

I did not address this third issue in 'The Common Mind', mainly because I couldn't find any basis for arguing either for or against it. But in recent years I have come to think that there are important considerations that argue against and much of my work has been given to exploring these, partly in single-authored writings and partly in collaboration with Christian List; we are currently writing a book on group agency.

Singularism has traditionally been rejected amongst a number of continental schools of philosophy but the argument against has sometimes romanticized corporate entities and this has often caused revulsion in other circles. Thus, under the influence of Popper and Hayek, English-speaking philosophy in the twentieth century generally adopted the view that ascribing beliefs or desires or values or actions to group or corporate entities is mere metaphor; that it is literally false or that it serves, at best, as an indirect way of ascribing intentional attitudes and actions to members or majorities within those groups. But this, it turns out, can't be right and it was the realization that it can't be right that led me, despite embracing individualism, to reject singularism as well as atomism.

Think of a mind as a system of attitudes, in particular a system of intentional attitudes like belief and desire. And think of an intentional agent as an entity within which such attitudes operate more or less consistently to a more or less consistent effect. Or, if the agent is reflective, think of it as an entity that can at least acknowledge the claims of consistency and recognize their relevance. The question as to whether group agents have minds of their own comes down, then, to the question as to whether there is a distinct system of attitudes formed and enacted by such a group or whether everything it does — everything done in its name by members and officers — can be fully explained without postulating such a system.

The usual singularist argument is that the attitude of a group on any proposition, be it an attitude or belief or desire or what-

11. Philip Pettit

ever, is bound to be a function of the attitudes of members, usually a majoritarian function; to say that the group believes or desires something, so the usual line goes, will be to say just that a majority of its members do so. But this, it turns out, cannot be right. For if a group is to behave consistently, or be responsive to the demands of consistency, then its attitudes on relevant propositions cannot generally be a function of the corresponding attitudes of individuals on those propositions.

That the attitudes of a group cannot be a majoritarian function of the attitudes of members can be shown by the existence of what I came to call discursive dilemmas; these are a generalized version of what are known as doctrinal paradoxes in the law and economics literature. A discursive dilemma shows that a number of people can have consistent attitudes, say beliefs, in respect of a set of connected propositions, while the majority set of views is inconsistent. Suppose A, B and C each have the views given in the following matrix on the connected issues of whether p, whether q, and whether $p\&q$. A and B and C will be individually consistent but the majority set of views will be inconsistent.

	p?	p?	$p\&q$?
A	Yes	No	No
B	No	Yes	No
C	Yes	Yes	Yes
Majority	Yes	Yes	No

The existence of such discursive dilemmas shows that Hobbes, Locke and Rousseau were mistaken to think that a group could do its business as a center of agency just by acting on the majority attitudes of its members. A group like this would have to act, impossibly, on the judgment that p, that q but that not $-p\&q$. If this group is to act in pursuit of various ends, on the basis of common judgments, then the members will have to decide to reject the majority view on at least one of the three propositions. The members will have to enact a system of attitudes that is not a majoritarian function of the corresponding attitudes among members. Thus they may decide to act on the judgment that $p\&q$, despite the fact that a majority of members reject that judgment.

Nor is this all. It also turns out that if a group is to be suitably consistent in the attitudes it forms and enacts on connected propositions, then on some propositions those attitudes cannot be guaranteed to be functionally related in any way, majoritarian or otherwise, to individual attitudes—not, at least, under plausible assumptions. The crucial assumptions are that the group follows a procedure of attitude – formation that works no matter what the attitudes of individuals are and that the group attitudes turn on the attitudes of more than one member. There are now a set of abstract results that bear on this matter, beginning with a result that Christian List and I published in 2002.

The upshot of all of this, by intuitive criteria, is that singularism is false and that there are groups with minds of their own. There are groups with systems of attitudes that are not a function of the corresponding attitudes of their members and that are in that sense independent. There is no mystery here, since it remains the case that anything the group comes to hold or do is determined by what the individual members hold and do. But there is surprise, for what the group comes to hold and do is determined in a complex manner that gives us ground for thinking of it as an independent agent.

The rejection of singularism has important implications for explanation in social science and for the investigation of incorporation, commercial and otherwise. It teaches lessons, for example, about how far group agents should be held responsible in their own right and about what it means for individuals to identify with group agents, giving those agents life in the way they form and enact its attitudes. These are issues with which I am currently engaged.

How do you conceive of the relation between the social sciences and the natural sciences?

I take a broadly Weberian view. Like natural science, social science may seek to build up bodies of data, identify salient patterns or laws that those data support and, if possible, construct theories that unify the laws and explain the generation of the data. But the divergence between the enterprises comes in the fact that the social scientist will have to show how any theory defended can be rendered consistent with the adoption of a stance, in traditional terminology, of *Verstehen*.

This is the stance I adopt on you, and invite you to adopt on me, when I take each of us to make sense to the other: to be someone who is conversable, in the sense of being disposed to recognize the force of commonly accepted reasons and, within the constraints of fallibility and weakness, to act as those reasons require. We cannot see others as persons, in the manner we take to be generally possible with human beings, unless they prove to be more or less conversable. We may be prepared to see some human beings as deranged and subject to forces and laws that make them unconversable; we may even be prepared to think that all human beings are subject to such derangement under limited circumstances. But we cannot we led to take such a view of all human beings all of the time; to do so would be to give up on the category of human persons. Hence we cannot take seriously a social science that claims to reveal that actually people are not conversable in the standard sense.

It is because of accepting this constraint on social science that I am an individualist: that is, someone who thinks that the regularities discovered or likely to be discovered in social science will not compromise our image of the human mind, as that is established in folk psychology and practice. Much of the philosophy of social science, as I see it, is given to showing how the most interesting of results can be squared, often despite appearances, with the assumption that human beings are conversable. Much of my own work on individualism in 'The Common Mind' is meant to explain why the existence of structural social regularities, of the kind that deeply impressed Durkheim, are consistent with continuing to view human beings as generally conversable.

What is the most important contribution that philosophy has made to the social sciences?

It should be no surprise, in light of what I said earlier, that I think the most important contribution of philosophy to the social sciences is to humanize them. This involves guarding against the assumption that the images of the human being developed in those sciences undermine the image we hold to in dealing with one another, as persons among persons. The philosophy of social science, at its best, displays a full appreciation of the range and depth of the findings and speculations of social science, while establishing how they can make sense from within the conversable stance.

This humanizing role is implemented on two fronts in contemporary philosophy of social science. On the one hand, such philosophy tries to show how the high-level social theories of sociology and anthropology can have validity, even supporting distinctively functional explanations of social pattern, without committing us to the view that the human subject is just a pawn of social forces, an epiphenomenon on the social stage. And on the other it attempts to explain how many austere, rational-choice accounts of social behavior may have application without committing us to the view that the human subject is nothing more than a machine for calculating and implementing the demands of an egoistical utility function.

Inevitably, some philosophical discussions of social science will have to be deeply critical, arguing for the weakness of the doctrines in question. But philosophers should be loathe to make allegations, having no empirical grounds on which to defend them: not at least, so far as they are just philosophers. Hence my view that their role should be to interpret the social sciences, and in particular to try to make sense of how certain patterns of explanation can work well, consistently with the conversable image that regulates our relationships as persons.

Which topics in the philosophy of social science will, and which should, receive more attention than in the past?

It is hard to address this question, as it is hard to address the others, without being somewhat self-referential, even narcissistic. Building on the comments just made, I would like to identify two key ideas with which I have worked myself in pushing the humanizing role of the philosophy of social science on the two fronts mentioned: one, with respect to structural theory, as we might call it, the other with respect to rational-choice theory. The ideas are those, respectively, of program explanation and virtual control.

Consider structural regularities to the effect that urbanization leads to secularization or decreasing employment to increasing crime. Social laws of these kinds are sometimes held to give credence to the idea that much of what happens in social life is determined behind people's backs, and determined in such a way that people do not have the autonomy as intentional subjects — 'autonomy' is more or less equivalent to 'conversability' — that common sense imputes. How to vindicate the humanistic image of persons in face of this challenge: a challenge that assumed its most influential form in the nineteenth and early twentieth century? My favored answer is: by resort to the program model of higher-level explanation.

Consider a natural process in which water in a closed flask is brought to the boil and, as a consequence, the flask breaks. Let us assume that what happens in the process is that as the water boils — as the mean motion of the constituent molecules reaches a suitable level — it becomes likely to the point of near-inevitability that some molecule will have a position and momentum sufficient to break a molecular bond in the surface of the flask; and that this in fact happens, leading to the collapse of the flask. What causes the flask to break in such a case? At one level the molecule that actually triggers the break in the surface is the cause of the collapse. But the fact that the water is boiling is also causally relevant to the event. Its relevance consists in the fact that given its presence — given that the water is boiling — it is more or less inevitable that there will be some constituent molecule, maybe this, maybe that, that has a position and a momentum sufficient to induce a crack in the surface of the flask.

The relationship between the causally relevant temperature and the causally relevant molecule might be described in terms of a metaphor from computing, as Frank Jackson and I have argued. The higher-level event — the water being at boiling point — programs for the collapse of the flask, and the lower-level event implements that program by actually producing the break. This program architecture applies in many cases of higher-level and lower-level causation but perhaps the most salient locus of application is with structural, higher-level social laws and the patterns displayed in the conversable behavior of individual human beings.

Consider, for example, the structural law relating falling employment with rising crime. That this obtains does not mean that the aggregate fact of falling employment works behind people's backs in some way to generate a rise in crime. Rather what happens is that it programs for a rise in crime and does so precisely by making it likely that people will behave in a way that leads to that aggregate effect. The falling employment means that there will be an increasing number of people with a need or wish for more money than they can obtain legitimately and with greater opportunity, created by the leisure that unemployment ensures, to seek such money by illegitimate means. Small wonder, then, if falling employment leads to rising crime. And what holds in this case is likely to hold in all the other cases where structural or aggregate factors are identified in social science as the causes of structural or aggregate effects. The pattern need not generate any suggestion that people's psychology is sidelined or undermined, once the supporting architecture of causation has the program profile.

The other key idea I mentioned is that of virtual control. What it serves to explain is how rational choice explanations can often be very useful in social science, without people conforming to the image of *homo economicus*: without their displaying the austerely calculative and self-concerned mentality associated with that image.

Whenever a causal process leads to a certain sort of effect, we may say that it controls for that effect; in particular that it controls actively for the effect, via the causal connections that are operative. But effects are often controlled for by factors that do not or need not be causally active in this way; they may be the products of virtual as distinct from active controllers.

A controller for a type of effect E — call it C — will be virtual under the following conditions. The effect E is normally occasioned, not by C, but by some other factor, N (for normal). But in any case where N fails to produce E, then C steps into the breach and takes over the productive role. When C steps in like this, it actively controls for the appearance of E. But so far as it is there as a standby cause, ready to intervene on a need-to-act basis, it controls for the appearance of C even when it is not actively in charge. It is a virtual controller of the effect in question. It rides herd on the normal mechanism of production, ready to compensate for failures in the process or to readjust the process so that it no longer fails.

The idea of virtual control makes sense of how rational self-interest can be invoked to explain ordinary, non-calculative patterns of action, even among people who do not conform to *homo economicus*. Even when people operate under perfectly ordinary forces of habit and custom, they may be subject to the virtual control of self-interest. It may be, first, that if those forces serve a person ill, this is likely to become evident — the red lights will go on — and, second, that should the red lights go on, then self-interest will become an active force, rationally triggering suitable amendments. The fact that rational self-interest has such a virtual presence in people's psychology should be neither a surprise or a scandal; by received wisdom, after all, most of us would be corrupted by the ring of Gyges, acting for our own ends under the invisibility it bestows. And the virtual presence of rational self-interest is sufficient to vindicate many of the claims, particularly the more persuasive claims, of rational-choice theory.

This claim can be illustrated with a range of rational-choice theories, ranging from the theory that conventional behavior is rationally supported by the coordination it makes possible to the thesis that slave-holding lasted as long as it did in the United States because it was comparatively profitable, to the view that compliance with oppressive dictatorships is explained by the fact that potential rebels are afraid that others will seek to free ride on their efforts. Geoffrey Brennan and I argue in a book on 'The Economy of Esteem' that the idea of virtual may also make sense of how a desire for esteem can rationally shape people's lives and relationships, without any one of them being very explicitly focused on the avoidance of disesteem or the pursuit of positive esteem.

These two ideas of program explanation and virtual control do not merely help to save the conversable image of human beings from the implications of social-scientific research. They may also serve to guide such investigation, as illustrated by the work on the economy of esteem. This means that perhaps I should qualify the earlier claim that the main role of philosophy is to humanize social science, understanding it in such a way that the conversable image remains in place. The highest aspiration may be, not just to understand social science, but also to change it.

12

Alexander Rosenberg

The R. Taylor Cole Professor of Philosophy

Department of Philosophy and Center for Philosophy of Biology, Duke University, USA

How did you get interested in the philosophical aspects of the social sciences?

I began my university studies in physics, made a conceptual mistake which resulted in my changing to philosophy. By the time I had corrected the mistake, reading Hume's Treatise, it was too late to return to physics. At that time in my university all degree candidates were required to take two economics classes. As a person of the left I had no interest in bourgeois economics, so I avoided these two classes as long as possible, and finally petitioned the committee on student requirements to be allowed to take only one on the pretext that that I wanted to take more philosophy classes. After only a few days of being introduced to microeconomics, I realized that this subject held great fascination for me and was a fertile topic for the philosopher of science.

I went to a philosophy graduate program in a university with an even stronger economics department, where I took many classes. When my philosophy department chairman suggested I write a thesis on the philosophy of economics, I took it as an order. It was the luckiest break of my professional life. Here was a subject on which no one in the latter day philosophy of science had written almost anything, and in which there were important papers and books by economists like Robbins, Machlup, and Friedman. Writing a dissertation in a "green field" not only freed me from tracking pervious work, but made whatever I wrote obligatory for others to refute once the field got going.

Of course given the period in which I began working on this subject, the late 60s, it was natural to adopt an empiricist post-Positivist agenda in the philosophy of economics, and to apply recipes and analyses drawn from positivists and their students to economics. In the UK it was the period during which Winch's *Idea of a Social Science*, and the other volumes in the tide of little red books (in Donald Davidson's words) were making philosophical waves, expounding a philosophy of science they attributed to the later Wittgenstein ("In psychology there are experimental results and conceptual confusions"). So, these two threads formed the line of my thesis. It was also the beginning of the Nixon stagflation in the US, in which the Keynesian consensus was unraveling. So it was also natural to ask the question why economics was not as predictively powerful and therefore scientifically successful as established theories in natural science.

In my later work the post-positivism and the Wittgensteinian view receded as targets of philosophical interest. But the question about predictive power and the "cognitive status" of microeconomics stayed with me.

Which social sciences do you consider particularly interesting or challenging from a philosophical point of view?

For me the enduring questions of the philosophy of social science are now continuous with the major issues of the philosophy of biology. Humans, *Homo sapiens* is s biological species, and the human sciences must all be compartments of biology. Since in biology we have the one and only right theory to explain adaptations, their emergence, persistence and improvement, we must perforce have the same advantage in social science. Despite the resistance of many social scientists, I think that Darwin's theory of natural selection is the only possible explanation of adaptation, that human affairs are all matters of local adaptation of something – a lineage, a group, a meme – and so we need to reconfigure social science to accord with the explanatory centrality of Darwinism. This conception has nothing much to do with nativism, evolutionary psychology, sociobiology or other substantive theses about the predominance of genetically encoded dispositions and traits over learned ones. On my view, blind variation and natural selection operate at many levels employing many replicators and interactors, and that so long as we identify the variables social scientists deal with as adaptations, we are obliged to find Darwinian style explanations for their emergence, persistence and improvement. Tracking these down is the most interesting task for the philosopher and the social scientist.

How do you conceive of the relation between the social sciences and the natural sciences?

My answer to the previous question should make it clear that I consider the social sciences to be biological sciences, and to act accordingly. This means in particular that the shape of their theories will have to follow the shape of theory in biology, that there will be no real laws, and a large number of local regularities, that the means/ends economy of human behavior, action, institutions, will have to be explained adaptationally, as noted above.

What is the most important contribution that philosophy has made to the social sciences?

I wish I could say that it has helped economics find a methodological prescription that it could actually live with, or that it is responsible for economics giving up the lip service it long paid to Popper while articulating a theory without treating it as falsifiable.

Probably the most significant "contribution" philosophy has made to social science has been negative and regressive. Its debates have provided antiempirical social scientists with arguments to defend their versions of social science from serious scrutiny. Compare for example Deirdre McCloskey's *The Rhetoric of Economics* or *If You are So Smart*, not to mention the vast tide of books protecting themselves by wrapping the mantle of Kuhn, *Structure of Scientific Revolutions* tightly around their discipline.

So, what has philosophy done for social science that has been of benefit to it? I think it has provided a great deal of conceptual clarification of particular research questions, most of them too specific to be visible to many philosophers or social scientists. If there is one visible concrete contribution that philosophers have made to the social sciences in recent years it must be the account of causation reflected in the work of Glymour, Spirtes and Scheines, as adopted by Judea Pearl, and expounded for economics by Kevin Hoover.

Which topics in the philosophy of social science will, and which should, receive more attention than in the past?

David Braybrooke used to illustrate the reflexive and unpredictable character of the human sciences with a quote from Louis Armstrong: When asked where jazz was heading, he said, "If I knew the answer to that, I'd be there already." I think we must say the same both about what social science will receive more attention in the future than the past, and what topic in philosophy of social science will do so as well.

13

David-Hillel Ruben

Professor of Philosophy

Birkbeck, University of London

Director and Professor of Philosophy

NYU in London, UK

The way in which I conceive of the philosophy of social science is, of course, in large measure a reflection of how I conceive of philosophical method more generally. My formative philosophical training in the early 1960's was in analytic philosophy. (This seems an excellent moment to pay tribute to my late teacher, Willis Doney, who was perhaps the single most formative philosophical influence on me.) I have always tried to make its (and his) standards of rigour and precision my own.

Although I was not explicitly conscious of it at the time, the philosophical method I imbibed was 'aprioristic' in at least one sense. I thought then, and I think now, that the main contribution philosophy can make to any field - science, religion, art, etc. - is to uncover the specifically philosophical questions that arise in that field, often but not always without the practitioners themselves being aware of those questions, to clarify what those questions are, and to try and make advances by discussing those questions in standardly philosophical ways. Philosophers 'ratiocinate' about the selected field, they apply philosophy to it, but they don't extract much from the target discipline in terms of method or even substance. Philosophers bring something to the special sciences that the latter need; they do not derive a great deal from them.

The questions that arise are typically metaphysical (sometimes, specifically ontological) and epistemological or methodological. I think of the philosophy of social science (and other 'philosophies of' as well), in large measure as applied metaphysics and epistemology. The standard techniques of those areas of philosophy can therefore be brought to the questions of social science. Philosophy's contribution is to help those sciences by using its own methods and assumptions for their benefit. I appreciate that to many this will sound philosophically arrogant. Whatever it sounds like, I think it is true.

The 1960's was important to me for another reason. Like many American university students affected in some way or another by the Viet Nam War, my philosophical interests began to extend to political philosophy. In particular, I became an admirer of the social thought of Karl Marx. I thought then, and think now, that Marx's system is an elegant and at least initially convincing analysis of society, how it is to be understood and how it changes. Of course, I have in the intervening 40 years found much in that system with which to disagree, but the intellectual power of the theory still fascinates me, and has set many of the issues in the understanding of society that I have subsequently addressed. I was never very interested in Marx's political ideas per se. Debates about whether or not there is a Marxian theory of justice, for instance, never interested me. The closest I came to an interest in that side of Marx was really methodological: how did Marx envisage the interplay between (what we would call) factual and evaluative statements?

What I found fascinating was the way in which the Marxian system puts together ideas in epistemology and metaphysics, sometimes borrowing from Hegel (and, via Hegel, the Greeks) and sometimes distancing itself from him. In my first book, *Marxism and Materialism*, I argued that Marx subscribed to a realist ontology. Marx's use of the term 'historical materialist' could easily confuse the reader, since I think that what he meant by that latter expression is what we would call 'realism'. Marx poses for us, in a way which we can easily appreciate today, problems in the understanding of materialism (or physicalism). Marx wanted to avoid the dangers of both Hegelian idealism and reductionist materialism. Marx said very little about epistemology, but I argued in that book that the attempt to graft an idealist epistemology and theory of truth onto his corpus, as some of the 'humanist' Marxists were fond of doing at the time, had no support in the texts themselves.

In studying Marx, I had been struck by the apparently contradictory Marxian views that, on the one hand, there was something wrong with Platonic/Hegelian ways of thinking which hypostasised abstractions, and, on the other, that there was something wrong with conceiving society in a methodological individualistic way. If society was not just a lot of individuals who are related in certain ways, then isn't society an irreducible entity in its own right? Why isn't society then a hypostasised abstraction after all? I therefore thought that questions of methodological individualism and holism, old chestnuts that they may be, still needed addressing. What sort of conception of society could avoid both the Scylla of reduction of society to individuals and the Charybdis of unwarranted abstraction? I believe that reconciliation is possible but the view that emerges must be far more subtle and nuanced than standard statements of either holism or individualism in the social sciences. So I wrote *The Metaphysics of the Social World*, in order to try and work through some of these issues in the ontology of social science.

I am hardly the first to have remarked on the near total absence of the category of action, and the absence of any explicit discussion of action, for the rather long period of philosophy between Aristotle (or perhaps the Medieval Aristotelians) and Hegel. Discussions about action sprang to life again in the 1960's and 1970's in the writings of philosophers of history, and then moved into the philosophical mainstream. Marx addresses the problem of action (quaintly called 'praxis' by him) in a way which I believe would not be sympathetic to the contemporary Davidsonian orthodoxy on action (he would have regarded the Davidsonian view as unacceptably reductionist and merely the 'vulgar materialism' that he identified in Hobbes and some of the French Encyclopaedists). So the intriguing question is: what kind of alternative understanding of action would a modern Marx (or Aristotle, for that matter) be willing to accept? For example, I don't give much philosophical hope to the idea that a person's acting on reasons can be well understood in terms of the same causality that natural science uses. Realist metaphysics sets the outlines or parameters, within which the other two questions, holism and action, can be discussed. In my *Action and its Explanation*, I turned to some of these issues.

The view that emerges from all of this looks something like this: society (by 'society', I mean to include such items as: actions, agents, rationality, warrant and justifiability, freedom, social objects, and so on) is where mind and matter meet. But society cannot be understood as decomposable into mind + matter. The social is something like an emergent reality. It is not a derivative idea; it is an unanalysable primitive, a conception which takes its place as primitive, alongside mind and matter. There is a wonderful quotation from Vladimir Medem, the Jewish Socialist leader in pre-revolutionary Russia: 'an individual in Russia was composed of three parts: a body, a soul, and a passport.' The view might well be called social realist: it asserts, against idealism, the irreducibility of matter. It is not so-called vulgar materialist, because it asserts the irreducibility of the social and the mental (this is probably what Marx meant by 'historical materialism'). Within that framework, the problems of action and social wholes can be tackled afresh. Of course, I don't mean to suggest that Marx would have welcomed being a dualist. He was struggling to articulate, in my view, a non-reductionist type of materialism that allowed scope for the reality, the irreducibility-but not the independence-of the mental and the social.

I believe that some action is free. I have often tried to think the thought that no one is free. I can just about make myself believe it about other persons but I cannot think this coherently about myself. If I were not free in any of my actions, then I cannot have freely come to the conclusion (because 'coming to a conclusion' is the name of a mental act) that I am not free. But if I know or even believe that I have not freely come to that conclusion, then I can have no reason to believe that it is something I am justified in believing at all. So I could believe that I am not free but I could never have any reason to think that my belief is anything more than an unjustifiable fact of nature.

Of course, many hold that some action is free in a soft determinist sense. Soft determinists think that freedom and causation (of the natural science type) are compatible. I find the thought that (some? all?) action is not caused in that sense a very attractive idea. Freedom and rationality, in my view, rule out causation. I don't find any so-called soft-determinist accounts of that freedom (or rationality, or acting for a reason, or justification, or warrant) at all convincing. No account of the causation of our action will leave us any better off than are puppets controlled and manipulated from the outside. Maybe we are only puppets but that is simply not consistent with our being free or rational, (sometimes) holders of justified beliefs. And since I hold that we sometimes act freely, act for reasons, hold justified beliefs, it follows that at least some of our actions, including our mental acts like deciding, judging, and so on, are not caused.

If all this is so, then there can't be any causal laws that apply to our actions. Indeed, I think that that is right. There are generalisations that one can make about human action, but these are what I would call 'generalisations from rationality'. This conception bears some affinity to what Michael Scriven called, a long time ago, 'truisms'. If I offer people a choice between accepting £10 or £20, no strings whatever attached, I can predict correctly that each person will accept £20. I can even produce a generalisation in some sense about this human behaviour. But it does not follow that there is any causation going on, or that the generalisation is an exceptionless or probabilistic law of nature. The generalisation rests on what we would expect rational agents to do, and that does not presuppose any causation at all. Rational choice theory need not presuppose anything about laws, causes, and so on.

What is the relationship between the natural and social sciences? The slogan, 'the natural sciences and social sciences are (or are not) the same', seems unfinished. One wants to add: 'in which respects'? There may be many ways in which they are similar. Clearly, from what I have said above, I believe that there are some important ways in which they seem not to be. Of course, to address this question, one must have a reasonably accurate view of what natural science is like, and I am not confident that my view is anything more than a nineteenth century caricature of natural science. (My *Explaining Explanation* does address issues of explanation in natural science though.) On the other hand, I have clearer views about what social science (or better, in order not to beg the question, social study) is like, so I am happy to leave to those whose understanding of natural science is deeper than mine to assess the level of difference between them, if what I say about social science is true.

The overall position I took above was that there are a set of concepts-action, acting for a reason, justification, freedom-which are distinctive to the social study of persons and societies, and which are not reducible to any of the concepts of natural science. One should not jump too quickly from views about concept irreducibility to views about the irreducibility of things themselves, for example, the things falling under those concepts. As philosophers use to say, formal mode and material mode questions are not always equivalent questions. For example, from the fact that physical object concepts cannot be reduced without remainder to phenomenological concepts, it does not immediately follow that physical objects themselves are not just sets of sense data. But with that caveat in mind, I would argue that in social science the irreducibility is both conceptual and ontological.

But what do we explain in social science? Explanation of individual action is by reasons for action, at least in the standard case, and what I have already said has the clearest implications for action explanation, although there are many actions I do for which I have no reason (pacing a train platform, kissing a picture of someone). If this is explanation by reasons and reasons for action are not amongst the causes of that action (as I believe), it follows that action explanation is not a species of causal explanation. But other things need explaining too in social science: whole patterns of behaviour (but maybe that is just an extension of action explanation, maybe not); social structures; social development and change. The list is hardly meant to be exhaustive. Each of these raises its own specific issues and must be addressed in that specificity. From the fact that action explanation is not by causes, it does not immediately follow that explanations of social change and development (especially if these social phenomena cannot be given individualistic reductions) are not causal either. Almost all available philosophical energy relating to explanation in the social sciences has been expended on individual action explanation. We need a good catalogue of the types of explananda that we find in the social science, and an account of the sort of explanation that explaining each type requires. Some of the work by Philip Pettit and Frank Jackson on programme explanation makes a start on this issue; much more needs to be done.

A number of issues in the philosophy of social science seem to me to be grossly under-discussed. The theory of the early Marx was universalistic in the sense that his primary focus was on people as part of humanity. Each of us is an individual, and also a member of a species. As his theory developed, he saw individuals in class societies as members of classes, but the goal of socialism was to do away with classes and allow individuals to once again assert their shared humanity with all of humankind. What Marx ignored was the lasting importance that individuals attribute to groupings less extensive than the whole of humankind. The ideas of traditions and cultures, their continuities and discontinuities over time, are an important set of ideas for understanding the modern world. I feel their impact in a deeply personal way. I have always felt my identity to be shaped in essential ways by my membership in two traditions: the philosophical tradition and the Jewish tradition. So for me, understanding traditions is also, at the same time, self-understanding.

There are many questions about traditions and cultures that

need answering, and the philosophical literature in these areas is almost non-existent. First, a number of closely related ideas need distinguishing: traditions, cultures, ideologies, systems of belief, movements, ways of life, practices, world views, schools of thought, and so on. If we focus for a moment just on the idea of a tradition, one question is simply: what is a tradition? What are the necessary and sufficient conditions for something's being a tradition? Even just with this one idea, it is not at all clear that everything rightly called 'a tradition' is so-called in the same sense. But, pending further clarification of the idea of a tradition, the question posed is the synchronic identification question.

Then there are diachronic identity questions. People and groups literally kill one another over disputes concerning which group is the true successor to some earlier group, or who is carrying on the authentic tradition, and so on. Who are the true successors to Mohammed, the Sunnis or the Shiites? Who is carrying on the tradition of the early church, the Protestants or the Catholics? A splendid pioneer in this area, whose writings have not been sufficiently taken up and developed, is W.B.Gallie on the topic of essentially contested concepts. Not enough work has been done in these areas: I think we need a philosophy of anthropology, which focuses on the ideas of tradition and culture, an area of social science much neglected by philosophers.

In the view of social science I adumbrated above, concepts such as rationality and justification are centre stage. Rationality must have some connection with the idea of rules, so issues about rules and rule-governed behaviour need addressing by philosophers of social science. Perhaps we became tired of rules, given their over-discussion and over-use by Wittgensteinians in the 1950's and 1960's. More recent work on rule-following has mostly been focussed on answering the sceptical questions that arise from Wittgenstein's writing. But we still do not have a fully satisfying account of what it is to obey a rule, as far as I am aware. Nor, come to that, do we have much of a grasp on the concept of rationality itself, beyond the minimalist idea of maximisation that one finds in accounts of Rational Economic Man. Early works by Amartya Sen (in his splendid 'Rational Fools'), Charles Taylor and others pose many of the right questions and even give some interesting answers about rationality, but all this needs further refinement. Richer accounts of rationality than the ones located within the utilitarian/welfare economics writers on the one hand and within the Kantian tradition on the other (with its emphasis on consistency) need elaborating.

Rules raise other issues too. There is also the issue of the importance of rules in society. An article that made an enormous impression on me was Stephen Lukes' comparison of Marx and Durkheim on rules. Does the imposition of rules on individuals alienate them, or does the absence of rules in the lives of individuals bring about anomie? Buried in this dispute are competing theories of human nature, of what constitutes human well-being. This dispute seems to straddle the moral/empirical divide, relying on both questions of substantive fact and moral theory. Part of my own problems with Marx concerns this point. The Durkheim view is plausible: that persons without rule-governed structures that confront them with categorical demands are unhappy in some way (one need not agree with his particular thesis about suicide, in order to accept the general insight). Marx's views on human liberation and emancipation, with the hindsight afforded by the twentieth (and now twenty-first) century, strike me as naïve and simplistic.

Laws are a particular kind of rule. Before the philosophy of law took a wrong turn into the intellectual cul-de-sac of critical legal studies and post-modernism, some great classics were produced in this field. H.L.A. Hart's *The Concept of Law* will prove to be, I think, one of the important philosophical texts of the twentieth century. I think that many in the philosophy of law have enhanced our understanding of rules via their serious and insightful work in that area, and yet philosophers of social science have not contributed to the debates they raise. Hart's (and other jurisprudentialists') work makes a contribution to certain methodological issues in the philosophy of social science, and not only to normative questions in ethics and social philosophy. One central debate-again, an old chestnut, but very much unresolved-is just this: what are the criteria for a rule being a valid law? Are those criteria solely factual or must they include a normative element? Answering these questions will lead us to understand whether, or in what sense, there can be a science of law and perhaps even a science of society.

Having raised the issue of facts and values above, let me now say something explicitly about it. One of the most-I am not sure quite what word to use here-murky areas in the philosophy of social science is the place of values in social science. John Searle made an important contribution to this issue some time ago, making it a part of the philosophy of social science, with his idea of institutional facts. Was he successful in 'bridging' the fact/value dichotomy in this way? I don't believe he was successful, but it is not easy to say why this is so, if indeed it is so, and I don't know of a compelling piece of philosophy that shows this.

Everyone, as far as I know, agrees with banalities like Weber's belief that value commitments can bring people to select one area of social study rather than another for investigation. But once we move beyond the banalities, the murk sets in. Is the language of social science value-free or can it be made to be so? What exactly is it for a language to be value-free? Do we have a workable criterion for assessing this? There are probably several connected ideas of a value, all tending to be run together in these discussions. The truth of a statement often has to be assessed by evaluating the evidence on which it is based. Is that in any way relevant to the idea we want to pinpoint in this debate?

Let's assume, for the sake of argument, that the language of social science is the ordinary, non-technical language of the man or woman on the Clapham omnibus. THAT language is clearly not value-neutral or value-free: concepts, some of which I have already used, such as alienation, anomie, rational, exploitative, slavery, harassment, deprivation, illness, freedom, and well-being, are not value neutral. Perhaps one could invent some technical language whose analogue concepts are free of those value implications. That language would have far less expressive power than the concepts we in fact use. So in part, the issue revolves around whether ordinary concepts and language are appropriate for social science, or whether social science requires the invention of a technical language, much as it does in chemistry, physics, biology, and so on. Some have tried to tread that last path-I am thinking here of Felix Oppenheim's work on freedom almost a half century ago, and much work in the Journal of Unified Science between the wars in Germany and Austria. None of this work, as far as I know, is thought to have been in the least successful. Assuming, then, that no such technical language is available to us, the answer to the question of values in social science seems to me so obvious that I wonder if I might not have missed something altogether straightforward.

When I think about the current state of affairs in the philosophy of social science, I have an overwhelming feeling of how little has been done and how much more work awaits us. I am also very aware of the limited number of questions that have attracted sustained, serious attention. Certainly, action is probably the greatest net receiver of attention, although here the prevailing Davidsonian orthodoxy has been somewhat stifling (this should not be exaggerated though, as there are some very important exceptions to this comment). Quite a few issues-rules, philosophy of law, human nature, culture and traditions, rationality-have simply faded from favour after an earlier period of being on the philosophical centre-stage. I think that, as always in philosophy, what are most needed are a lively sense of imagination, a willingness to create new answers to old questions and to invent new questions for the asking, and a sense that where the answer seems to us most obvious, there is a lurking orthodoxy waiting to be overturned.

14

John R. Searle

Slusser Professor of Philosophy

University of California at Berkeley, USA

How did you get interested in the philosophical aspects of the social sciences?

In a sense, I have always been interested in the philosophical aspects of the social sciences. Even before I knew there were such things as the "social sciences" or "philosophical aspects" I was worried about questions such as, Do we genuinely have free will? and I think those have a direct bearing on the social sciences. When I seriously began to study philosophy it seemed to me there was a terrific ontological problem about the social sciences, and that worried me for many years, I do not even know how long, until finally I wrote a book about it, published in 1995, *The Construction of Social Reality*. The question is this, How can there be a class of objective facts in the world that are only facts because we believe them to exist? Money, national governments, private property, marriage, and football games are all what they are only under certain descriptions. But they only exist under those descriptions because we believe that they exist. Some groups of people could go through the same set of physical movements, but it would not be a football game unless they had the appropriate mental attitudes. This is the philosophical aspect of the social sciences that most interests me, and the question can be put very succinctly in the form of a paradox, How can there be epistemically objective facts that are ontologically subjective? "Epistemically objective" means that we can find out as an objective matter of fact whether or not such and such is a piece of money or counterfeit or whether or not George Bush is now president. But the facts that we discover contain an element of human subjectivity, because they depend on the attitudes that people have, and in that sense they are ontologically subjective.

Which social sciences do you consider particularly interesting or challenging from a philosophical point of view?

Let us arbitrarily define the social sciences as consisting of Sociology, History, Political Science, Cultural Anthropology, Economics, and Psychology or at least Social Psychology. Now, it seems to me, all of these disciplines are interesting from a philosophical point of view, though not all in the same way. The social sciences are all, in their different ways, about human behavior. The behavior in question typically consists of intentional actions. These actions are typically performed for a reason or reasons. That means that the social sciences deal with a certain aspect of human rationality. Human rationality presupposes free will. This means that in a very deep sense the social sciences are all about human free will.

All of these reflections have enormous implications for the social sciences. Some of the various social sciences I have mentioned have specific philosophical problems associated with them. So, for example, the form of economics that I was taught as an undergraduate in Oxford had an extremely impoverished conception of human rationality. Rationality, for the economist, was always a matter of maximizing utility, or some such, and this is an extremely impoverished theory of rationality. I have attacked it in a book called *Rationality in Action*. But not all social sciences make that mistake, though some make other mistakes. There is a kind of endemic cultural relativism that seems to infect Cultural Anthropology, and this is a philosophical mistake though I have never bothered to answer it publicly, except indirectly with my attacks on post-modernism. The problem with Political Science, for example, is that it tends to lapse into a kind of journalism, because the research often does not have a serious enough philosophical grounding to ask the philosophically interesting questions about political power and the political processes. Sometimes a "great work" in political science will often only be valid for a few decades.

How do you conceive of the relation between the social sciences and the natural sciences?

The social sciences are radically different in their form of explanation from the natural sciences. There are some exceptions from what I am about to say, but in general we can say that the form of explanation in the social sciences is done by giving intentional causation. We specify the intentional states – beliefs, desires, and so on – that motivate human actions. But this is quite different from the natural sciences where the notion of intentional causation does not figure at all. There are some exceptions to this in some branches of social sciences. We can give explanations in terms of climate changes, or diseases, killing populations, but for the most part explanations of social phenomena are given in terms of intentional causation, such as beliefs, desires, hopes and fears. But intentional causation presupposes rationality, and rationality presupposes free will. So, though we live in one world, which consists ultimately of physical particles as described by the physicists, the level of explanation that we seek in the social sciences is radically different from the sort of explanation that we receive in the physical sciences.

What is the most important contribution that philosophy has made to the social sciences?

It seems to me too early to answer this question because I believe the sort of thing that I consider to be serious philosophical reflections on the social sciences is fairly recent, and indeed is practiced by rather few philosophers. The two questions that seem to me most needing an answer are first the question of social ontology, What exactly is the mode of existence of social and institutional entities, such as governments, political parties, football teams, universities, nation-states, and religious movements? and secondly, What explanatory mechanisms do we need to use in explaining the behavior of these phenomena, whether the behavior of individual actors or the behavior of whole social movements? And there, as I said in response to earlier questions, we need a different type of explanatory apparatus than we have in the natural sciences. All of this needs very deep philosophical reflection.

Which topics in the philosophy of social science will, and which should, receive more attention than in the past?

I cannot really make a prediction about what is actually going to be happen, but from what I said earlier it should be clear that I think what is truly essential is that we should get clear about the ontology of social phenomena. We should be clear about the mode of existence of political parties, football teams, nation-states, religious movements and marriages. And we should also be clear about the nature of the explanatory apparatus that we use to explain the behavior of these phenomena; both the behavior of individual actors within the social structures and the behavior of entire social movements.

My general point is that I think the area of social reality, the subject matter of the social sciences, is wide open for philosophical research, and I wish to encourage it.

15

Raimo Tuomela

Professor of Practical Philosophy
University of Helsinki, Finland

The following interview questions were presented by the editors for me to answer:

1. How did you get interested in the philosophical aspects of social science?

2. Which social science do you consider particularly interesting, challenging, or problematic from a philosophical point of view?

3. How do you conceive of the relation between the social sciences and the natural sciences?

4. What is the most important contribution that philosophy has made to the social sciences?

5. Which topics in the philosophy of social science will, and which should, receive more attention than in the past?

My answers to 1–4 will be very short, but I will give a relatively long answer to topic 5:

1. I studied philosophy, psychology, social psychology, sociology, and mathematics as a university student and graduated in psychology. Since then I have studied economics and international politics on my own. I have a Ph.D. degree in theoretical philosophy from the University of Helsinki (Finland) and another Ph.D. degree, also in philosophy, from Stanford University (USA). I got interested in the philosophical aspects of social science probably because I have since school times liked abstract thinking and radical thinking in the old sense deriving from the Latin word "radix"

that means root and because I had been interested in psychology and social psychology even during my school years.

2. Social psychology, sociology, economics, and political study are all theoretically and philosophically interesting in their own way. Thus social psychology basically is concerned with "jointness" and "sharedness" as well as social groups and acting in them. The term "meso level" is sometimes used for these phenomena. Sociology is concerned with the macro level consisting of various kinds of groups, especially large groups, and social structures. Both the meso level and the macro level involve problems related to the individualism-holism debate in the social sciences. Economics as we now have it is very individualistic and atomistic, and it is concerned mainly with strategic acting and utility maximization. Some theorists in economics have claimed that it is the fundamental social science from which the other social sciences can be built. This claim is questionable from a philosophical point of view.

I find international politics interesting because it deals with cooperation and conflicts between large groups (nations and states). There are theoretically new features involved here that are not to be found on the micro or meso (or jointness) levels. However, I should say that my most central interests concern the foundations of the social world in toto, in particular problems related to what an adequate underlying conceptual framework for the study of the social world should be and problems related to social ontology.

3. The basic problem here is how to view human agents and the products of their activities in relation to the extensional, basically non-intentional entities that such basic natural sciences as physics deal with. Thus it can be said that it is human intentionality (and collective intentionality) and what it embraces (e.g. questions of interpretation of mental states and actions, morality, etc.) that present problems for a unified picture of the world. It must be kept in mind, though, that the natural sciences are taken to include also biology and therefore e.g. such animals as apes. Thus, given that e.g. chimpanzees and some other mammals have rudimentary intentionality the central question is how to account for intentionality in general, more or less independently which biological species are capable of having it. Do intentionality and feelings and sensations pose a big problem for the creation of an integrated naturalistic view of the world? My view is that intentionality and language use indeed are central features of human beings but that a suitable naturalistic, but still non-reductive account of them still can be given; and many preliminary accounts

already are available, although lots of work remains to be done.

4. At least one important general contribution is that philosophical research has emphasized the importance of the conceptual framework of persons and agency. This framework conceives of human beings as thinking (thus intentional) and acting beings that can be (morally) responsible for their activities. By their collective acceptance and creation human groups, viz. members of groups collectively, can create social institutions. Thus institutions are collective artifacts, something that in principle have been totally "made up" by the group members. It can thus be said that the collective intentionality (joint intentions, goals, beliefs, etc.) that collective acceptance can be taken to involve is most central for understanding the social world. Accordingly, social theories should use concepts representing various phenomena of collective intentionality, as such social notions as institutions, groups, organizations, norms, and what can be built out of them do conceptually rely on collective intentionality. The strongest form of collective intentionality is based on the full-blown "we-perspective", viz. on the group members taking themselves to be parts of a "we" and of their accordingly acting as proper group members rather than as private persons.

5. I will below clarify the we-perspective approach mentioned in 4. This also indicates what my approach to the conceptual and metaphysical foundations of the social sciences is (for a comprehensive account, see Tuomela, 2007):

The we-perspective can be argued to be central for the correct understanding and explanation of the social world—over and above the I-mode perspective. My technical explicate for the *full* we-perspective (group perspective) is the we-mode. Roughly speaking, the we-mode is concerned with group-involving states and processes that the group itself has at least partly conceptually and ontologically constructed for itself (e.g., the group may simply take as its goal—"our" goal for the members—to build a bridge). Acting as a group member in the we-mode sense *constitutively* involves acting for a collectively constructed *group reason*—the group gives a group member reasons to think, "emote," and act in certain ways. For instance, the group's constitutive goals, values, and beliefs provide such group reasons. In contrast, the I-mode is concerned *only* with "private" personal and interpersonal reasons and relations as well as with groups involving such ingredients. Group reasons in a weaker sense may contingently be involved. In general, the we-mode and the I-mode complement each other.

The we-mode is indispensable for understanding the most basic group-level social concepts. For instance, social institutions require special thickly collectively constructed contents (e.g., "Euros are *our* money"), and this arguably falls beyond the conceptual resources of the I-mode. The important divide here is between a group thinking and acting as one agent versus some agents acting and interacting, perhaps in concert, in pursuit of their (possibly shared) private goals. Group reasons (we-mode reasons) and I-mode reasons (private reasons) for acting and having attitudes thus are clearly of a different kind. Accordingly, only the we-mode can properly account for the generality that the group level involves with respect to group members (change of membership, future members, etc.) and the kind of (partial) depersonalization that group life involves.

It is often useful to view a group as an agent capable of acting as a unit. Thus it can be taken to accept views, form intentions, act, and be responsible. However, it is not an extra agent over and above the group members. When a group acts, its members must act as group members. In a sense, one can thus redescribe the group's functioning and acting at the group member level, in terms of the group members' functioning in appropriate ways as group members. This is basically we-mode activity. It follows from the idea of a group acting or functioning as one agent that the members ought to function appropriately. They can be said to be necessarily "in the same boat," "stand or fall together," or share a "common fate." They satisfy the so-called Collectivity Condition. Formulated for the special case of goal satisfaction this condition necessarily connects the members as follows: Necessarily (as based on group construction of a goal as the group's goal), the goal is satisfied for a member if and only if it is satisfied for all members. The Collectivity Condition is a central constitutive element of the we-mode.

While also the I-mode we-perspective (and I-mode collective intentionality) exists, the we-mode we-perspective is holistic and richer—e.g., in that it involves the Collectivity Condition and the group reason requirement, both of which are based on collective construction in the group. We-mode collective intentionality (equaling the we-mode we-perspective) basically amounts to thinking and acting fully as group member. The I-mode case is partly different, for there is also I-mode collective intentionality that does not involve the I-mode we-perspective. Notice, too, that people often if not typically act both for we-mode and I-mode reasons, even on the same occasion.

The we-mode relies on the central distinction between thinking and acting as a group member (reflecting the group level) versus as a private person and it builds on the former of these notions. The second element, over and above thinking and acting as a group member, conceptually involved in the we-mode is collective commitment. In order for a group member to act as a group member, it must thus be required that she be committed (bound) to performing actions that further the group's constitutive goals, beliefs, standards, etc. and other matters that the group is pursuing. Indeed, the members should be collectively committed, viz., committed as group members, to participate in group activities. Their collective commitment involves that they are also "socially" committed (viz., directedly committed to other group members) to each other to act in the right group ways. Collective commitment has two basic, intertwined roles here. First, it "glues" the members together around an ethos. This gives the foundation for the unity and identity of the group. Secondly, collective commitment serves to give *joint* authority to the group members to pursue group-related action. They can and must in their own thinking and acting take into account that the group members are collectively committed to the group ethos and to the group members and that they are jointly responsible for promoting the ethos. Every group member is accountable not only to himself for his participatory action but also to the other members. All this shows how group unity as formed by collective commitment to the ethos relates to action as a group member.

In view of the above, it can be said that the we-mode is constituted by two elements, a content element (the mentioned kinds of constitutive contents) and an action-related element, viz., collective commitment. To illustrate, we consider a two-person case in which a goal (or intention, belief, etc.) is collectively accepted (constructed) and held by two persons, you and me. The case involves two elements:

(i) G is *our* goal, where "our" satisfies the aforementioned Collectivity Condition.

(ii) We (you and I together) are collectively committed to goal G.

I claim that (i) and (ii) give the intuitive "rock bottom" of the we-mode. Actually, (ii) is part and parcel of (i) and entailed by it. The participants being collectively committed to goal G involves that they are committed to doing their parts of their joint action concerned with their achieving G. The joint goal that they here have constructed for their group serves as their reason for their performing their parts. The notion of a joint goal satisfies the Collectivity Condition. Due to its being satisfied, the notion of "we" is not reducible to the conjunction "you and I" although it entails it.

The we-mode elements (i) and (ii) can intuitively be viewed as translations of group level descriptions of the following kinds:

(i') group g's goal is G (where g has you and me as its sole members)

(ii') group g is committed to goal G.

(i') and (ii') can be regarded as equivalent. Hence also (i) and (ii) are seen to be equivalent from our present group-level perspective. The present point applies, mutatis mutandis, also to intentions, beliefs, and other voluntary attitudes.

The fullest notion of the we-mode requires that the group members collectively accept the group's constitutive and other goals and beliefs, etc., for group use and do it only in their action but are also at least disposed to accept them in the reflective and reflexive sense that the specific constellation of goals, values, beliefs, etc., indeed are the group's goals and beliefs, all this being publicly available knowledge in the group.

Let me briefly point out how we-mode thinking and acting helps to clarify *cooperation*, which surely is a central social notion (see Tuomela, 2007, Chapter 7 for discussion and arguments). Here is a short summary of the basic distinctive benefits of we-mode cooperation as contrasted with I-mode cooperation: The we-mode is central for understanding the core concept of cooperation—involving acting together towards a shared goal. The we-mode approach is based on shared goals, intentions, beliefs, etc. and thus gives uniform motivating group-reasons to all participants. As a result it creates order on both the group level and the member level than I-mode approaches (mainly because of the collective commitment involved). It gives more stability, persistence, and often also more flexibility to cooperative action. Part of the extra connectedness involved here is due to shared social capital and respect-based trust. The we-mode approach can be more rewarding even in a utility-maximizing sense. In relation to empirical cooperation research the following comment can be made. The we–mode involves that the group takes responsible for the group members' actions as group members and thus it involves relevantly monitoring and controlling other group members and sanctioning, if need be. Hence it entails "strong reciprocity" that has recently been argued to be central in empirical research of cooperation. Strong reciprocity means that the participants in cooperation not only should be disposed to act cooperatively and reciprocate on the "first-order" level but should also be disposed to punish and sanction those who defect, and this includes also those who defect in the sense of failing to punish defectors. Mere reciprocity is not sufficient to explain cooperation, strong reciprocity is needed according to this research. The we-mode approach simply entails strong reciprocity: the group members functioning properly as group members, thus in the we-mode, are disposed to cooperate with fellow group members, given that they cooperate; and the monitoring and controlling they are engaged in involves also advising, correcting, and sanctioning those who do not appropriately cooperate.

The we-mode in principle avoids collective action dilemmas (e.g. PD) at the ingroup level; however, the real dilemma here rather is whether to be a "we-moder" or an "I-moder". Because of being based on group authority, the we-mode often required for group change and crisis management, e.g. in the case of natural catastrophes, intergroup conflicts, etc.

Current social science has begun to incorporate the kinds of social elements that the we-mode approach contains. A case in point is *new institutionalism,* which in recent years has been doing that at least in the fields of sociology, political science, and economics. What still seems to be missing, though is a systematic use of group notions and we-notions in the clarification of social activities. Both the idea of a group (from the point of view of "us") as an explanatory and justificatory element and the more prevalent idea of interpersonal connections are needed for successful social theorizing, and the former notion has not to my knowledge found its way to social science research.

As seen, the we-perspective approach relies on the conception of human beings as persons in the sense of the "framework of agency" that assumes that (normal) persons are thinking, experiencing, feeling, and acting beings capable of communication, cooperation, and of following rules and norms. There are important evolutionary considerations related to this. One is the phylogenetic fact that the homo species has had at least the general capacity for both we-mode and I-mode attitudes and action presumably for hundreds of thousands of years, although not refined by linguistic skills until perhaps one or two hundred thousand years ago. The other is the fact or at least claim that children acquire a rudimentary capacity for shared intentionality towards the end of their first year of life and for we-mode collective intentionality starting as early as in their second year. In this process the child learns to make-believe and pretend. This capacity is central, e.g., for a person's understanding the notion of institutional status. The disposition to have collectively intentional, viz., we-mode, thoughts and to act in the we-mode seems to be a coevolutionary adaptation (viz., based on a genetic and cultural evolutionary mechanism and history). The precise content of the collectively intentional mental state is nevertheless culturally and socially determined. It has been argued on the basis of experimental results, that the most basic feature that distinguishes human beings as a species from higher animals such as chimpanzees is the humans' capacity and motivation to have, and act on, collectively intentional states that probably coevolved long ago, perhaps in connection to the emergence of modern *homo sapiens*.

References

Tuomela, R. (2007), *The Philosophy of Sociality: The Shared Point of View*, Oxford University Press, Oxford

About the Editors

Diego Ríos is currently lecturer at the Department of Philosophy and Economics at the University of Witten/Herdecke, Germany. He has been teaching at the Universities of Versailles, Düsseldorf, Cologne, and Hertfordshire.

Christoph Schmidt-Petri is currently lecturer in practical philosophy at the University of Leipzig. He received a PhD in philosophy from the LSE in 2005 and since then has held positions at the Universities of Witten/Herdecke, Glasgow, and Saarbrücken.

Index

a priori, 1, 5, 74, 79, 108
Abbott, A., 75, 77, 83
absolutism, 10
action, 7, 8, 33, 37, 48–51, 53, 54, 65–67, 72, 81, 82, 85, 86, 88, 90, 92, 95, 96, 100, 106, 107, 112, 117, 123, 127, 130–134, 137, 140, 141, 144, 146–149
Adams, J., 83
analytical Marxism, 80, 81
Andreski, S., 103, 113
anthropology, 5, 33, 35, 47, 55, 96, 104, 121, 135, 140
anything goes, 10, 11
Aristotle, 31, 36, 37, 131
Armstrong, L., 127
Aron, R., 90
Asian, 81, 82
Austin, J.L., 2

Balliol College, 89
Balogh, T., 89
Barkow, J., 111, 113
Barnard College, 57
Barnes, B., 9
Bartlett, Sir F., 4–7
Bayesian Nets, 106
Beck, G., 32
Beethoven, L. van, 51
Bergson, H., 32
Berlin, Sir I., 89
biology, iv, 33, 35, 36, 41, 72, 74, 76, 77, 85, 95, 110, 111, 113, 114, 125–127, 137, 144
Blalock, H., 60
Blatt, J., 34
Bloch, M., 38
Block, N., 110, 113
Bottomore, T., 90
Brandt, R., 98
Bratman, M., 51, 52, 55
Braudel, F., 38
Braybrooke, D., 127
Brennan, G., 124
Brodbeck, M., 79
Burge, T., 107, 113

Cambridge Controversy, 57, 58
Capital, 58, 80
capitalism, 81
capitalist, 80
Carruthers, P., 98, 99
Carson, T.L., 98, 100
Cartwright, N., 74, 77, 83, 105, 113
causation, 60, 61, 65, 67, 71–73, 87, 88, 97, 122, 123, 127, 132, 133, 141
chemistry, 64, 85, 110, 137
chestnut, 82, 131, 136
China, 81, 84
Chomsky, N., 79
class, 37, 40, 91, 134, 139
cognition, 5, 6, 10, 79, 85, 93–95, 97, 99

cognitive science, 76, 77, 113, 114
Cohen, G., 81, 91, 96, 99
Cohen, P., 81
Coleman, J.S., 33, 92, 99
collective agency, 92, 108
commitment, 50–52, 62, 63, 91, 93, 137, 147, 148
common knowledge, 48–51
comparative historical sociology, 83
complexity, 4, 79, 108
Comte, A., 93
conceptual analysis, 2, 6, 13, 54
continental philosophy, 86
contingency, 84, 85, 92
convention, 1, 5, 8, 35, 36, 48–50, 52, 74, 78, 103, 124
Cook, J.W., 98, 99
cooperation, 67, 85, 144, 148, 149
Cosmides. L., 111, 113
Cuba, 84

Dahrendorf, R., 90
Darwin, C., 76, 110, 126
Davidson, D., 65, 107, 113, 126, 131, 138
Debreu, G., 41
Descartes, R., 31
Dobb, M., 79
Doney, W., 129
Durkheim, E., 36, 47, 49, 53, 54, 74, 79, 90, 96–98, 100, 120, 136

Eatwell, J., 57, 58
economics, 34, 35, 38, 39, 41, 47, 55, 57–60, 62, 69–71, 74, 75, 77, 82, 83, 87, 89, 96, 104, 109, 118, 125–127, 135, 140, 143, 144, 149, 151
Economics and Philosophy (journal), 60
Eddington, Sir A., 32
Edge, D.O., 9
Elster, J., 81, 92
equality, 40, 60, 93
evolutionary psychology, 35, 111, 113, 126
experiment, 5, 10, 32, 39, 64, 67, 75–77, 83, 93, 105, 106
experimental, 2–4, 6, 7, 39, 66, 72, 74–76, 98, 113, 126, 150
explanation, iv, 9, 59, 60, 63, 65, 66, 70, 72–77, 80, 81, 84, 85, 87–89, 91, 92, 94, 95, 98, 100, 104, 106, 107, 113, 119, 121, 123, 124, 126, 131, 133, 134, 140, 141, 145
externalism, 107

Febvre, L., 38
Feyerabend, P., 104
Findlay, R., 58
Finley, M.I., 79
Flew, A., 2
Fodor, J., 106, 107, 109, 110, 113
Freud, S., 31
Friedman, M., 34, 59, 60, 125

Gallie, W.B., 135
Geertz, C., 86, 92, 99
Gellner, E., 90, 100
Gentile, G., 32
Germani, G., 33
global warming, 67

Glymour, C., 67, 127
Goldstone, J., 83
Goodman, N., 69
Guala, F., 74, 77

Hart, H.L.A., 136
Harvard University, 99, 100, 114
Hauser, M.D., 98–100
Hayek, F.A. von, 96, 117
Hegel, G.W.F., 31, 32, 130, 131
Hercowitz, M., 103
hermeneutics, 35, 36, 86
Hesse, M., 3, 7
history, 3, 5, 6, 9, 10, 33, 35, 62, 63, 79–81, 84, 90, 91, 95, 97, 99, 103, 104, 112, 116, 131, 140, 150
Hobsbawm, E., 79
holism, 33, 40, 87, 104, 131, 144
holistic, 146
Hollis, M., 91, 100
homo economicus, 59, 123
Hoover, K., 127
human action, 85, 92, 132, 141
human condition, 10
Hume, D., 1–4, 8, 10, 12, 61, 62, 105, 125

idealism, 12, 130, 132
ideology, 8, 36, 40
intentional explanation, 106, 107
intentionality, 6, 66, 95, 97, 144–146, 150
intentions, 37, 51, 53, 54, 106, 145, 146, 148
interdisciplinarity, 83
interdisciplinary, 9
Iran, 84

Jackson, F., 122, 134
Jeans, Sir J., 32
justice, 40, 83, 93, 112, 130

Kant, I., 32, 37, 90, 135
Keynes, J.M., 34, 126
Kincaid, H., 70, 77, 78, 87
Klein, V., 103
knowledge, 1, 4, 5, 9–11, 32, 47–51, 60, 62, 64, 67, 72, 73, 80–82, 84, 86, 87, 92, 95–97, 106, 110, 114, 117, 148, 149
knowledge, empirical, 10
knowledge, inductive, 10
Kohli, A., 82
Kuhn, T.S., 104, 127

Lévy-Bruhl, L., 98
Ladd, J., 98
Lakatos, I., 104
large-N, 84
Laurence, S., 99
Leibniz, G.W., 31
Levi, I., 58
Levy, N., 98, 100
Lewis, D., 48, 52, 55, 74, 78
liberty, iii, 40
Lieberson, S., 85
List, C., 117, 119
Lopasic, A., 103

Macdonald, G., 110, 113, 115
Machlup, F., 125
MacIntyre, A., 91
macrosociology, 75
Mameli, M., 112
Mannheim, K., 103
market, 8, 38, 40, 60, 75
Marx, K., 31, 38, 63, 66, 79–81, 86, 91, 96, 99, 130–132, 134, 136

Marxism, 32, 80, 81, 86, 130
materialism, 38, 91, 96, 130–132
mathematics, 1, 33, 34, 36, 103, 112, 143
Mathiez, A., 33
Maxwell, J.C., 59
McCloskey, D., 127
Mead, M., 98
Medem, V., 132
Mellor, H., 58
meme, 126
mental states, 106, 107, 144, 150
Merton, R., 33
microfoundations, 88
Milgram, S., 105
Mill's methods, 103, 104
Mill, J.S., 59, 66, 103, 104, 106
MIT, 82, 114
Moody-Adams, M., 98, 100
Moore Jr., B., 83
Morgenbesser, S., 58
Morgenstern, O., 34
Moser, P.K., 98, 100
Myers, C.S., 4
Myrdal, G., 89, 94, 100

naturalism, 35, 36, 62, 63, 66, 69, 72, 86, 114
naturalistic, 5, 6, 10, 84, 144
Newton Smith, B., 91
Newton, I., 59

obligation, 50, 51, 95
ontology, iv, 35, 38, 64, 76, 77, 81, 86, 87, 93, 100, 116, 130, 131, 141, 144
Osborne, M., 58
Osborne, R., 31
Oviedo, 34

Parsons, T., 96
perception, 3, 7
perfect competition, 59
Perkins, D., 82
physics, 2, 8, 31–34, 36, 38, 41, 47, 74, 75, 85, 101, 109, 110, 113, 125, 129–131, 137, 144
Pierson, P., 83
plasticity, 85
Plato, 31, 32, 130
political philosophy, 49, 50, 52, 79, 116, 130
political science, 55, 84, 96, 104, 140, 149
political sociology, 35, 71, 75
Popper, K., 95, 104, 117, 127
positivism, 73, 82, 86, 93, 126
psychology, iv, 3–6, 31–33, 35, 39, 55, 64, 76, 79, 85, 89, 98, 104, 106, 111, 113, 116, 120, 122, 123, 126, 140, 143, 144
Putnam, H., 107, 114

quantum mechanics, 33
Quine, W.V.O., 69

rationality, 12, 60, 74, 80, 86, 90–92, 95, 100, 101, 131–133, 135, 138, 140, 141
Rawls, J., 79, 93, 99
realism, 104, 130
reason, iii, 2, 4, 10, 49, 50, 52, 58, 61–63, 66, 70, 72, 80, 82, 87, 99, 106–108, 110, 112, 116, 130, 132–134, 140, 145, 146, 148
reasons, iii, 40, 54, 65, 66, 71, 73, 84, 94, 95, 110,

111, 116, 120, 131, 132, 134, 140, 145–148
reductionism, iv, 36, 87, 110
reductionist, 76, 92, 130–132
relativism, iv, 10–12, 87, 91, 92, 99, 104, 140
religion, 38, 129
responsibility, 7, 94
risk, 40
ritual, 97
Rivers, W.H.R., 4
Robbins, L., 125
Robinson, J., 34
Rousseau, J.-J., 52, 118
rule-following, 97, 115, 116, 135
Russell, Bertrand, 31
Russell, Bill, 103
Ryle, G., 2, 3

Samuels, R., 111, 114
Samuelson, P., 34, 89
Sartre, J.P., 115
Scheines, R., 67, 127
Scott, J., 81
Scriven, M., 133
Searle, J., 51, 52, 90, 92, 93, 95, 97, 100, 136
Segal, G., 107, 114
Sellars, W., 51
Sen, A., 92, 93, 96, 135
Shapin, S., 9
Simmel, G., 47, 79
Simon, H.A., 60, 85
Skinner, G.W., 81
Skocpol, T., 83
Skyrms, B., 65
small-N, 83
Smith, A., 66
Soboul, A., 79
social capital, 148

social psychology, 4, 39, 104, 140, 143, 144
Social Science Research Council, 82
socialism, 134
socialist, 96, 132
sociobiology, 35, 95, 111, 126
sociologism, 36
sociology, 5, 8, 9, 33–36, 38, 47, 48, 53, 71, 73, 75, 83, 84, 86, 90, 91, 95, 96, 98, 100, 103, 104, 110, 120, 140, 143, 144, 149
sociology of knowledge, 9, 10
Spirtes, P., 67, 127
Sraffa, P., 58
Stanford University, 143
Stein, H., 58
Steinmetz, G., 83
Strawson, P.F., 2
Streeten, P., 89, 100
Strong Programme, 36
structuralism, 115
Sucre, M.G., 34

Taylor, C., 49, 73, 74, 78, 89, 91, 94, 95, 100, 101, 125, 135
terrorism, 37, 67
Thelen, K., 83
theory, 8, 33–36, 40, 41, 48–50, 53, 57–60, 62, 65, 66, 72, 74, 76, 79–84, 86, 88, 89, 91–93, 96, 99, 100, 104, 116, 119, 121, 123, 124, 126, 127, 130, 133, 134, 136, 140
theory of meaning, 66
Thompson, E.P., 69, 79
Tilly, C., 83
Timmer, P., 82

Tomasello, M., 55
Tooby, J., 111, 113
Trinity College, Cambridge, 3

Universidad de Buenos Aires, 32, 33
University of Bradford, 115
University of Edinburgh, 1, 9
University of Helsinki, 143
University of Illinois, 79
University of Keele, 1
University of Pennsylvania, 32
University of Reading, 103
utilitarianism, 93, 96
utility, 34, 40, 121, 140, 144, 148

Verstehen, 37, 97, 119
Vidal, P., 38
Viet Nam, 130

Waddington, Peter (Tank), 103
Wallerstein, I., 79
Watson, A.J., 3, 4, 31
Weber, M., 37, 47, 48, 53, 54, 63, 74, 79, 94, 96, 97, 137
Westermarck, E., 98
Wilson, B.R., 90, 101
Wilson, E.O., 111, 114
Winch, P., 48, 90, 92, 93, 101, 126
Wittgenstein, L., 1–3, 12, 65, 97, 115, 116, 126, 135
Wong, R.B., 83
Woodward, J., 83
Wylie, A., 74, 78

www.ingramcontent.com/pod-product-compliance
Lightning Source LLC
Chambersburg PA
CBHW022011160426
43197CB00007B/388